# Analytical Performance Modeling for Computer Systems

Third Edition

# Synthesis Lectures on Computer Science

Analytical Performance Modeling for Computer Systems
Y.C. Tay
2010

The Theory of Timed I/O Automata
Dilsun K. Kaynar, Nancy Lynch, Roberto Segala, and Frits Vaandrager
2006

Analytical Performance Modeling for Computer Systems, Third Edition
Y.C. Tay

ISBN: 978-3-031-00675-3   paperback
ISBN: 978-3-031-01803-9   ebook
ISBN: 978-3-031-00071-3   hard

DOI: 10.1007/978-3-031-01803-9

A Publication in the Springer series
*SYNTHESIS LECTURES ON COMPUTER SCIENCE*

Lecture #10
Series ISSN
Synthesis Lectures on Computer Science
Print 1932-1228   Electronic 1932-1686

# Analytical Performance Modeling for Computer Systems

## Third Edition

Y.C. Tay
National University of Singapore

*SYNTHESIS LECTURES ON COMPUTER SCIENCE #10*

# ABSTRACT

This book is an introduction to analytical performance modeling for computer systems, i.e., writing equations to describe their performance behavior. It is accessible to readers who have taken college-level courses in calculus and probability, networking, and operating systems.

This is not a training manual for becoming an expert performance analyst. Rather, the objective is to help the reader construct simple models for analyzing and understanding the systems that they are interested in.

Describing a complicated system abstractly with mathematical equations requires a careful choice of assumptions and approximations. They make the model tractable, but they must not remove essential characteristics of the system, nor introduce spurious properties.

To help the reader understand the choices and their implications, this book discusses the analytical models for 40 research papers. These papers cover a broad range of topics: GPUs and disks, routers and crawling, databases and multimedia, worms and wireless, multicore and cloud, security and energy, etc. An appendix provides many questions for readers to exercise their understanding of the models in these papers.

# KEYWORDS

computer system performance, analytical modeling techniques, simulation, experimental validation, Markov chains, queueing systems, fluid approximation, transient analysis

# Contents

# Preface

In writing this book, I have in mind the student, engineer or researcher who

(a) is interested in the performance of some particular computer system, and
(b) wants to analytically model that behavior, but
(c) does not intend to become an expert in performance analysis.

For network systems, the literature has numerous examples of analytical performance models; evidently, this research community has found such models useful. This book would have served its purpose if it can help the researchers in other communities (hardware architecture, operating systems, programming languages, database management, etc.) add a modeling chapter to a thesis or a similar section in a paper.

There is a common perception that performance modeling requires a lot of queueing theory. This is not so. Queueing systems are used in this book only as an expository device for a coherent presentation. The concepts (e.g., open/closed in Chapter 1, residual life in Chapter 3, flow equivalence in Chapter 6, stability in Chapter 8), techniques (e.g., Markov chains in Chapter 4, Average Value Approximation and fluid approximation in Chapter 7, equilibrium decomposition in Chapter 8), and results (Little's Law in Chapter 1, effect of variance on response time in Chapter 2, PASTA in Chapter 5, bottleneck analysis in Chapter 6) are applicable to more than just queueing systems.

In the first edition, I chose 20 papers to illustrate the ideas in the book. The papers were as follows.

**(1)** *StreamJoins*

J. Kang, J. F. Naughton, and S. Viglas. Evaluating window joins over unbounded streams. In *Proc. IEEE Int. Conf. on Data Engineering (ICDE)*, 341–352, March 2003.

**(2)** *SleepingDisks*

Q. Zhu, Z. Chen, L. Tan, Y. Zhou, K. Keeton, and J. Wilkes. Hibernator: helping disk arrays sleep through the winter. In *Proc. ACM Symp. Operating Systems Principles (SOSP)*, 39(5):177–190, October 2005.

**(3)** *GPRS*

G. Nogueira, B. Baynat, and P. Eisenmann. An analytical model for the dimensioning of a GPRS/EDGE network with a capacity constraint on a group of cells. In *Proc. ACM MOBICOM*, 215–227, August 2005.

**(4)** *InternetServices*

> B. Urgaonkar, G. Pacifici, P. Shenoy, M. Spreitzer, and A. Tantawi. Analytic modeling of multitier Internet applications. *ACM Trans. Web*, 1(1):2, 2007.

**(5)** *OpenClosed*

> B. Schroeder, A. Wierman, and M. Harchol-Balter. Open versus closed: a cautionary tale. In *Proc. Symp. Networked Systems Design and Implementation (NSDI)*, May 2006.

**(6)** *TCP*

> J. Padhye, V. Firoiu, D. Towsley, and J. Kurose. Modeling TCP throughput: a simple model and its empirical validation. In *Proc. ACM SIGCOMM*, 303–314, September 1998.

**(7)** *BitTorrent*

> D. Qiu and R. Srikant. Modeling and performance analysis of BitTorrent-like peer-to-peer networks. In *Proc. ACM SIGCOMM*, 367–378, 2004.

**(8)** *CodeRed*

> C. C. Zou, W. Gong, and D. Towsley. Code red worm propagation modeling and analysis. In *Proc. ACM Conf. Computer and Communications Security (CCS)*, 138–147, November 2002.

**(9)** *WirelessCapacity*

> P. Gupta and P.R. Kumar. The capacity of wireless networks. *IEEE Trans. on Information Theory*, 46(2):388–404, 2000.

**(10)** *802.11*

> Y. C. Tay and K. C. Chua. A capacity analysis for the IEEE 802.11 MAC protocol. *Wireless Networks*, 7(2):159–171, 2001.

**(11)** *MediaStreaming*

> Y.-C. Tu, J. Sun, and S. Prabhakar. Performance analysis of a hybrid media streaming system. In *Proc. ACM/SPIE Conf. on Multimedia Computing and Networking (MMCN)*, 69–82, January 2004.

**(12)** *StorageAvailability*

> E. Gabber, J. Fellin, M. Flaster, F. Gu, B. Hillyer, W. T. Ng, B. Özden, and E. A. M. Shriver. Starfish: highly-available block storage. In *Proc. USENIX Annual Tech. Conf.*, 151–163, June 2003.

**(13)** *TransactionalMemory*

> A. Heindl, G. Pokam, and A.-R. Adl-Tabatabai. An analytic model of optimistic software transactional memory. In *Proc. IEEE Int. Symp. on Performance Analysis of Systems and Software (ISPASS)*, 153–162, April 2009.

**(14)** *SensorNet*

R. C. Shah, S. Roy, S. Jain, and W. Brunette. Data mules: Modeling a three-tier architecture for sparse sensor networks. In *Proc. IEEE Workshop on Sensor Network Protocols and Applications*, 30–41, May 2003.

**(15)** *NetworkProcessor*

J. Lu and J. Wang. Analytical performance analysis of network processor-based application design. In *Proc. Int. Conf. Computer Communications and Networks*, 78–86, October 2006.

**(16)** *DatabaseSerializability*

P. A. Bernstein, A. Fekete, H. Guo, R. Ramakrishnan, and P. Tamma. Relaxed currency serializability for middle-tier caching and replication. In *Proc. ACM SIGMOD Int. Conf. Management of Data*, 599–610, June 2006.

**(17)** *NoC*

J. Kim, D. Park, C. Nicopoulos, N. Vijaykrishnan, and C. R. Das. Design and analysis of an NoC architecture from performance, reliability and energy perspective. In *Proc. ACM Symp. Architecture for Networking and Communications Systems (ANCS)*, 173–182, October 2005.

**(18)** *DistributedProtocols*

I. Gupta. On the design of distributed protocols from differential equations. In *Proc. ACM Symp. on Principles of Distributed Computing (PODC)*, 216–225, July 2004.

**(19)** *NonstationaryMix*

C. Stewart, T. Kelly, and A. Zhang. Exploiting nonstationarity for performance prediction. In *Proc. ACM EuroSys Conf.*, 31–44, March 2007.

**(20)** *SoftState*

J. C. S. Lui, V. Misra, and D. Rubenstein. On the robustness of soft state protocols. In *Proc. IEEE Int. Conf. Network Protocols (ICNP)*, 50–60, October 2004.

In the second edition, I added 10 papers (and exercises) to illustrate the concepts, techniques, and results in this book, and demonstrate breadth in their application. Several were chosen to show how an analytical model can offer insight not available through experiments. These 10 papers were as follows.

**(21)** *WebCrawler*

J. Cho and H. Garcia-Molina. The evolution of the web and implications for an incremental crawler. In *Proc. Int. Conf. on Very Large Data Bases (VLDB)*, 200–209, September 2000.

**(22)** *DatacenterAMP*

V. Gupta and R. Nathuji. Analyzing performance asymmetric multicore processors for latency sensitive datacenter applications. In *Proc. Int. Conf. Power Aware Computing and Systems*, 1–8, October 2010.

**(23)** *RouterBuffer*

G. Appenzeller, I. Keslassy, and N. McKeown. Sizing router buffers. In *Proc. ACM SIG-COMM*, 281–292, August 2004.

**(24)** *DependabilitySecurity*

K. S. Trivedi, D. S. Kim, A. Roy, and D. Medhi. Dependability and security models. In *Proc. Int. Workshop on Design of Reliable Communication Networks (DRCN)*, 11–20, October 2009.

**(25)** *PipelineParallelism*

A. Navarro, R. Asenjo, Si. Tabik, and C. Cascaval. Analytical modeling of pipeline parallelism. In *Proc. Int. Conf. on Parallel and Architectures and Compilation Techniques (PACT)*, 281–290, September 2009.

**(26)** *Roofline*

S. Williams, A. Waterman, and D. Patterson. Roofline: an insightful visual performance model for multicore architectures. *Commun. ACM*, 52(4):65–76, April 2009.

**(27)** *Gossip*

R. Bakhshi, D. Gavidia, W. Fokkink, and M. Steen. An analytical model of information dissemination for a gossip-based protocol. In *Proc. Int. Conf. Distributed Computing and Networking (ICDCN)*, 230–242, January 2009.

**(28)** *EpidemicRouting*

X. Zhang, G. Neglia, J. Kurose, and D. Towsley. Performance modeling of epidemic routing. *Computer Networks*, 51(10):2867–2891, July 2007.

**(29)** *P2PVoD*

B. Ran, D. G. Andersen, M. Kaminsky, and K. Papagiannaki. Balancing throughput, robustness, and in-order delivery in P2P VoD. In *Proc. CoNEXT*, 10:1–10:12, November 2010.

**(30)** *CloudTransactions*

D. Kossman, T. Kraska, and S. Loesing. An evaluation of alternative architectures for transaction processing in the cloud. In *Proc. ACM SIGMOD Int. Conf. Management of Data*, 579–590, June 2010.

For the third edition, I used similar criteria to select another 10 papers (with corresponding exercises).

**(31)** *SoftErrors*

A.A. Nair, S. Eyerman, L. Eeckhout, and L.K. John. A first-order mechanistic model for architectural vulnerability factor. In *Proc. IEEE Int. Symp. Computer Architecture (ISCA)*, 273–284, June 2012.

**(32)** *GPU*

J.-C. Huang, J.H. Lee, H. Kim, and H.-H. S. Lee. GPUMech: GPU performance modeling technique based on interval analysis. In *Proc. IEEE/ACM Int. Symp. Microarchitecture (MICRO)*, 268–279, December 2014.

**(33)** *ProactiveReplication*

A. Duminuco, E. Biersack, and T. En-Najjary. Proactive replication in distributed storage systems using machine availability Estimation. In *Proc. ACM CoNEXT Conf.*, 27:1–27:12, December 2007.

**(34)** *ServerEnergy*

B. Guenter, N. Jain, and C. Williams. Managing cost, performance, and reliability trade-offs for energy-aware server provisioning. In *Proc. IEEE INFOCOM*, 1332–1340, April 2011.

**(35)** *DatabaseScalability*

S. Elnikety, S. Dropsho, E. Cecchet, and W. Zwaenepoel. Predicting replicated database scalability from standalone database profiling. In *Proc. ACM EuroSys Conf.*, 303–316, April 2009.

**(36)** *PerformanceAssurance*

N. Roy, A. Dubey, A. Gokhale, and L. Dowdy. A capacity planning process for performance assurance of component-based distributed Systems. In *Proc. ACM/SPEC Int. Conf. Performance Engineering (ICPE)*, 259–270, September 2011.

**(37)** *CachingSystems*

V. Martina, M. Garetto, and E. Leonardi. A unified approach to the performance analysis of caching systems. In *Proc. IEEE INFOCOM*, 2040–2048, April 2014.

**(38)** *InformationDiffusion*

Y. Matsubara, Y. Sakurai, B.A. Prakash, L. Li and C. Faloutsos. Rise and fall patterns of information diffusion: model and implications. In *Proc. ACM KDD*, 6–14, August 2012.

**(39)** *MapReduce*

E. Vianna, G. Comarela, T. Pontes, J. Almeida, V. Almeida, K. Wilkinson, H. Kuno, and U. Dayal. Analytical performance models for MapReduce workloads. *Int. J. Parallel Programming* 41(4):495–525, August 2013.

**(40)** *ElasticScaling*

> D. Didona, P. Romano, S. Peluso, and F. Quaglia. Transactional Auto Scaler: Elastic scaling of in-memory transactional data grids. In *Proc. Int. Conf. Autonomic Computing (ICAC)*, 125–134, September 2012.

These papers were also chosen to demonstrate various aspects of performance modeling: applying the MVA algorithm to nonseparable queueing networks (*DatabaseScalability* and *PerformanceAssurance*), how to factor in the effect of precedence constraints on delays (*MapReduce*), unrolling a state transition diagram (*ServerEnergy*), how a model can be used to control a system (*ProactiveReplication*), the difference between whitebox and blackbox models (*SoftErrors* and *ElasticScaling*), how machine learning can help (*GPU* and *ElasticScaling*), and using a model to develop the science in system behavior (*CachingSystems* and *InformationDiffusion*).

But most important of all, I hope the readers who I have in mind can gain confidence from studying these examples, and see that they too can construct an analytical model for the performance behavior of their system.

Y.C. Tay
Kent Ridge
Westwood 2010
Palo Alto 2013
Taipei 2017
June 2018

# Acknowledgments

The drafts of this book were tested with courses at NUS (National University of Singapore), UCLA (University of California, Los Angeles) and NTU (National Taiwan University). Many thanks to my hosts Dick Muntz at UCLA and Yuh-Dauh Lyuu at NTU, and to the reviewers for their encouraging comments. I finished the second edition while on sabbatical at VMware; I thank the many colleagues there who made my visit so enjoyable and productive. Thanks too to Diane Cerra of Morgan & Claypool for her encouragement, and Clovis L. Tondo for his invaluable help with the typesetting.

Y.C. Tay
June 2018

# CHAPTER 0

# Preliminaries

This chapter presents some basic definitions and results that are frequently used in modeling computer performance. We will use *italics* when referring to an equation, figure, table, section, etc., in a cited paper.

While the Gaussian (or normal) distribution is most important in statistical analysis, the dominant distribution in performance analysis is exponential.

**Definition 0.1**
Let $\lambda$ be a positive real number. A continuous random variable $X$ has an **exponential distribution** with parameter $\lambda$ (denoted $X \sim \text{Exponential}(\lambda)$) if and only if

$$\text{Prob}(X \leq t) = \begin{cases} 1 - e^{-\lambda t} & \text{if } t \geq 0, \\ 0 & \text{if } t < 0. \end{cases}$$

□

This distribution is widely used in performance modeling, for which $t$ is usually time, so $\lambda$ is a **rate**.

**Lemma 0.2**
*If $X \sim \text{Exponential}(\lambda)$, then $EX = \frac{1}{\lambda}$ and $VarX = \frac{1}{\lambda^2}$.*  □

Note that, the higher the rate $\lambda$, the smaller the mean *EX* and variance *VarX*.

The exponential distribution is very simple. If it is used to model a performance metric (e.g., latency) for each component of a larger system, then these exponential distributions combine to give more complicated distributions for the system. In particular we have the following.

**Definition 0.3**
Let $X_1, \ldots, X_k$ be independent random variables, $X_i \sim \text{Exponential}(\lambda_i)$.
(a)    Suppose $k \geq 2$. The random variable $Y = X_1 + \cdots + X_k$ has a **Hypoexponential distribution**; it is called **Erlang-$k$** if $\lambda_1 = \cdots = \lambda_k$.
(b)    Suppose $\lambda_1, \ldots, \lambda_k$ are not all equal. Let $p_1, \ldots, p_k$ be real numbers, $0 < p_i < 1$ and $p_1 + \cdots + p_k = 1$. The random variable $Z$ such that $\text{Prob}(Z = X_i) = p_i$ for $i = 1, \ldots, k$ has a **Hyperexponential distribution**.  □

In other words, a series of exponential distributions gives a hypoexponentially distributed sum (see Fig. 1), whereas a choice from exponentially-distributed alternatives results in a hyperexponentially distributed result (see Fig. 2).

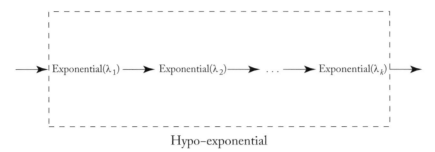

Hypo–exponential

Figure 1: Exponential distributions in series: hypoexponentially distributed sum.

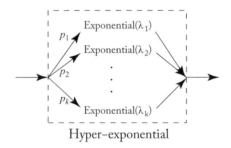

Hyper–exponential

Figure 2: Exponential distributions in parallel: hyperexponentially distributed choice.

We exclude $k = 1$ in (a) and $\lambda_1 = \cdots = \lambda_k$ in (b) because the distributions are then simply exponential.

For example, if we use an exponential distribution to model each server in *Fig. 2* in *InternetServices* [63], and we assume the delays are independent, then the latency for a client request would be hypoexponentially distributed. Similarly, the server delay for each tier in *Fig. 1* in *InternetServices* [63] would be hyperexponentially distributed. In reality, the service times for a request at two servers can be strongly correlated, so independence is violated. Nonetheless, such series/parallel construction of hypo/hyerexponential distributions is common in performance modeling.

In general, the mean and standard deviation of a distribution are different, and their ratio defines the following measure.

**Definition 0.4**
The **coefficient of variation** for a random variable $X$ is $C_X = \frac{\sqrt{VarX}}{EX}$.                    □

The names "hypoexponential" and "hyperexponential" are based on the coefficient of variation.

**Theorem 0.5**
*The coefficient of variation for hypoexponential, exponential, and hyperexponential distributions are less than, equal to and greater than 1, respectively.* □

Intuitively, for $X_1 + \cdots + X_k$, some $X_i$ value may be smaller than $EX_i$ and some $X_j$ value may be bigger than $EX_j$, so the effect is to reduce the total variation; for $X_1, \ldots, X_k$ in parallel, the differences among $EX_1, \ldots, EX_k$ increase the variation.

The exponential distribution has a discrete analog, namely the following.

**Definition 0.6**
Let $p$ be a real number, $0 < p < 1$. A discrete random variable $N$ has a **geometric distribution** with parameter $p$ (denoted $N \sim \text{Geometric}(p)$) if and only if

$$\text{Prob}(N = k) = (1 - p)^{k-1} p \qquad k = 1, 2, \ldots.$$

□

This distribution is what you would get if you perform independent trials, where each trial has a probability of "success," and $N$ counts the number of trials till the first "success." Thus, if $p$ is small, then we expect $N$ to be large.

**Lemma 0.7**
*If $N \sim \text{Geometric}(p)$, then $EN = \frac{1}{p}$ and $VarN = \frac{1-p}{p^2}$.* □

Here, note the similarity between Lemmas 0.2 and 0.7. In fact, these two distributions share the following property.

**Definition 0.8**
A continuous random variable $X$ is **memoryless** if and only if

$$\text{Prob}(X > s + t \mid X > s) = \text{Prob}(X > t) \quad \text{for any } s, t > 0;$$

a discrete random variable $N$ is **memoryless** if and only if

$$\text{Prob}(N > h + k \mid N \geq h) = \text{Prob}(N > k) \quad \text{for any } h, k \geq 0.$$

□

The memoryless property makes the exponential and geometric distributions unique.

**Theorem 0.9**

*A continuous random variable is memoryless if and only if it is exponentially distributed; a discrete random variable is memoryless if and only if it is geometrically distributed.*                    □

Thus, if the time $T$ that a Ph.D. student takes to graduate has a memoryless distribution, then after 4 years, the probability that she will take another 3 years or more is just $\mathrm{Prob}(T > 3)$, since $\mathrm{Prob}(T > 7 \mid T > 4) = \mathrm{Prob}(T > 3)$.

Similarly, if the number $N$ of babies it takes for a mother to get a boy is geometrically distributed, then after 5 girls, the probability the next baby is a boy is simply $\mathrm{Prob}(N = 1)$, since $\mathrm{Prob}(N = 6 \mid N \geq 5) = \mathrm{Prob}(N = 1)$.

It is clear that the memoryless property is a strong requirement. Nonetheless, it is widely used (often implicitly) in derivations as, otherwise, a complicated system would become analytically intractable. This is why the exponential and geometric distributions are so common in performance analysis.

In some cases, it may be possible to justify using the exponential distribution with the following.

**Theorem 0.10**    *(Renyi)*
*Suppose $X_1, \ldots, X_N$ are independent and identically distributed continuous random variables with $EX_i = m > 0$ and $N \sim \mathrm{Geometric}(p)$. Then*

$$X_1 + \cdots + X_N \sim \mathrm{Exponential}\left(\frac{p}{m}\right) \quad \text{approximately, for small } p.$$

□

Suppose we are modeling packet transmission over $N$ hops, where per-hop delays are independent. If the time for each hop is exponentially distributed and $N$ is a constant, then the end-to-end delay is hypoexponentially distributed; but if $N$ is geometrically distributed and the per-hop delays are identically distributed (whatever that distribution may be), then the end-to-end delay is—by Renyi's Theorem—exponentially distributed, approximately.

Renyi's Theorem seems to say that an exponential distribution naturally results from summing a large $(1/p)$ number of $X_i$ that are identically but arbitrarily distributed, rather like the Central Limit Theorem for a normal distribution. However, the theorem is actually not saying much, since we avoid one memoryless assumption (exponential) by assuming another one (geometric). We will return to Renyi's Theorem when we introduce Average Value Approximations (Sec. 7.1).

The exponential distribution is closely related to another discrete distribution.

**Definition 0.11**
A discrete random variable $N$ has the **Poisson distribution** with parameter $m$ (denoted $N \sim$

Poisson($m$)) if and only if

$$\mathrm{Prob}(N = k) = e^{-m} \frac{m^k}{k!} \quad \text{for } k = 0, 1, 2, \dots.$$

□

Its mean and variance are the same.

**Lemma 0.12**
*If $N \sim \mathrm{Poisson}(m)$, then $EN = VarN = m$.*  □

One way to see the relationship between the exponential and Poisson distributions is to consider **jobs** arriving at a **system**.

**Definition 0.13**
Suppose a system has arrivals at times $T_0, T_1, T_2, \dots$, where $T_0 < T_1 < T_2 < \dots$. If $X_i = T_i - T_{i-1}$ for $i = 1, 2, \dots$, then $X_i$ is called an **inter-arrival time**.  □

**Theorem 0.14**
*Assume inter-arrival times $X_1, X_2, \dots$ are independent and $X_i \sim \mathrm{Exponential}(\lambda)$ for $i = 1, 2, \dots$. Let $s$ and $t$ be positive real numbers, and $N_s(t)$ the number of arrivals between $s$ and $s + t$. Then $N_s(t) \sim \mathrm{Poisson}(\lambda t)$ for any $s$.*  □

This theorem leads to the following terminology.

**Definition 0.15**
Arrivals that have inter-arrival times that are independent and exponentially distributed with parameter $\lambda$ are called **Poisson arrivals**, with **arrival rate** $\lambda$.  □

Poisson arrivals are easy to work with, in the following sense.

**Theorem 0.16**
**(Additive Property)**
*Suppose a system has two independent streams of Poisson arrivals, with rates $\lambda_1$ and $\lambda_2$. Then they combine to give Poisson arrivals with rate $\lambda_1 + \lambda_2$.*
**(Split Property)**
*Suppose a system has Poisson arrivals at rate $\lambda$, and each arrival is independently labeled $\mathcal{L}$ with probability $p$. Then the arrivals that are labeled $\mathcal{L}$ are Poisson arrivals, with rate $p\lambda$.*  □

For example, if the requests leaving the load balancer in *Fig. 1* of *InternetServices* [63] have independent and exponentially distributed inter-departure times, and their choices of web server (Tier 1) are independent with constant probability, then the Split Property says each server gets Poisson arrivals.

Similarly, if the inter-departure times from each web server are independent and exponentially distributed, then the Additive Property implies that the request streams merge into Poisson arrivals at the Tier 2 dispatcher.

The crux of the theorem lies not in the resulting rates $\lambda_1 + \lambda_2$ and $p\lambda$ (which are intuitive), but in the resulting distributions. For example, if the input streams have constant inter-arrival times, then the combined streams would in general have non-constant inter-arrival times. Similarly, if the input stream has constant inter-arrival time, then a labeled substream will have nondeterministic inter-arrival time if $0 < p < 1$.

CHAPTER 1

# Concepts and Little's Law

We begin by introducing various concepts. We then state the single most important result in analytic performance modeling.

## 1.1 CONCEPTS

A **system** has multiple components. For example, the system in Fig. 6 of *SleepingDisks* [68] has a set of clients, a local area network, a database server, a storage area network, and a storage server. One of the first tasks in constructing a performance model is to decide which components to focus on, and at what granularity; a server, for example, can itself be viewed as a system with a storage cache and a set of disks.

The system has a **state** $S(t)$ at time $t$, which one may view as a vector of its component states, say $S(t) = \langle s_1(t), \ldots, s_m(t) \rangle$, where $m$ is the number of components and $s_i(t)$ is the state of component $i$. For example, $s_i(t)$ may just be the number of jobs at a server, or specify in greater detail the description of each job, the status of the jobs that are running, etc.

We are only interested in **dynamic** systems, for which $S(t)$ changes with time $t$. Parts of the system (e.g., the database) that are **static** (i.e., does not change with time) are usually omitted from $S(t)$.

Some of the changes in $S(t)$ may be **deterministic**. For example, the number of mobiles $N$ in a cell in *GPRS* [42] is fixed, and one can think of this as a departing mobile being deterministically replaced by an arriving mobile.

However, changes in any real system are usually caused by probabilistic events; for example, while the congestion window in *TCP* [43] may be reduced by half (deterministically), the reduction is induced probabilistically by the loss of a packet. The window size thus becomes a random variable that varies with time. This is an example of a **stochastic process**, i.e., a random-valued function $f(x)$, also called a **time series** if $x$ is time, or a **random field** if $x$ is space. For example, the node density for a network of randomly located wireless nodes in *WirelessCapacity* [21] is a random field.

Although $S(t)$ may be a stochastic process, one might seek to simplify its analysis by adopting a deterministic model. For instance, the worm propagation model in *CodeRed* [69] treats the number of infected hosts at any moment as deterministically governed by differential equations.

This study of worm propagation is an example of **transient analysis**, where one examines how $S(t)$ varies with time $t$. In general, transient analyses are difficult, and most performance models focus on the **steady state** $S_\infty = \lim\limits_{t \to \infty} S(t)$. For example, *BitTorrent* [44] starts by looking at the steady state for $\langle x(t), y(t) \rangle$, where $x(t)$ and $y(t)$ are the number of downloaders and seeds at time $t$. Note that $S_\infty$ may be a random variable; in that case, we are interested in the steady state distribution of $S_\infty$.

A performance model takes a set of **input parameters** that are of particular interest, and studies how their values affect system performance. For example, in *Table 1* of *802.11* [55], the input parameters are $n$, $W$, $m$, $T_{slot}$, $T_{SIFS}$, $T_{DIFS}$, $T_{payload}$, $T_{physical}$, and $T_{ACK}$.

As you can see, a model may have many input parameters, and some models may choose to hold certain parameters constant; e.g., the hash bucket size $B/|B|$ in *StreamJoins* [28] is fixed at 10.

The performance of a system is measured by some **performance metrics**, and an analytical model has the task of deriving these metrics from the input parameters. For example, the main performance measure in *OpenClosed* [46] is mean **response time** for the user, which measures the time lag between generation and completion of a job. Examples of other common performance measures include throughput (e.g., number of completed jobs per second) and utilization (e.g., percentage of busy time) for the system.

## 1.2    OPEN AND CLOSED SYSTEMS

Suppose we have a multiprogramming system with a processor and two disks, and we want to analyze how job completion times are affected by queueing for processor time and disk access. We can abstractly model the system as three queues, as shown in Fig. 1.1, where the arrows indicate job flow.

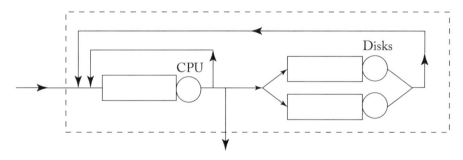

Figure 1.1: An abstract model of a computer system with a processor and two disks.

One of the first tasks in performance modeling is deciding how jobs arrive at the system. In an **open model**, jobs arrive at the system at some rate that is independent of system state, and leave when they are done—as illustrated in Fig. 1.2.

The input parameter to the model is the arrival rate $\lambda$, and performance measures—functions of $\lambda$—to be determined by the model may be average number of jobs in the system $N(\lambda)$, average

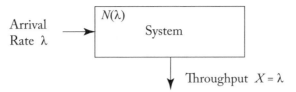

Figure 1.2: Open model: jobs arrive independently of system state. Performance measures are functions of $\lambda$.

time in the system $T(\lambda)$, utilization, etc. Note that, in steady state, the throughput $X$ would just be $\lambda$.

In a **closed model**, the system has $N$ jobs, and every completed job is immediately replaced by a new job, as illustrated in Fig. 1.3; the input parameter $N$ (sometimes called **population size**) is thus constant. The performance measures—functions of $N$—to be determined may be throughput $X(N)$, average time in the system $T(N)$, utilization, etc.

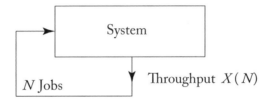

Figure 1.3: Closed model: a completing job is immediately replaced by an arriving job. Performance measures are functions of population $N$.

Metrics like throughput and utilization measure system performance, and are of interest to administrators. Human users perceive performance mostly through **response times**. If one is particularly interested in how these depend on the number of users, then it would be more appropriate to use an **interactive model**. This is a special case of a closed system, in which each job is generated by a user who, after that job completes, sends another job after some **think time**—see Fig. 1.4.

The input parameters are the number of users $N$—a constant, so the system is closed—and a think time $Z$ specified either as an average or with a density distribution (e.g., exponential). The central metric would be average response time. Other performance measures include throughput and utilization.

Think time is sometimes referred to as **sleep time**. However, it may be useful to give them different meanings. For example, consider a closed model for the web surfing behavior of a fixed number of users [59]. In each surfing **session**, a user may click one hyperlink after another, and "think time" would be the time between the completion of a download and the next click, whereas "sleep time" would be the time between the completion of a session and the next session.

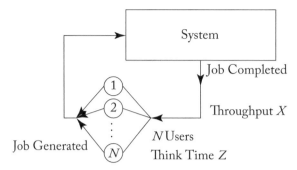

Figure 1.4: Interactive model: each job is generated by a user, with a think time between a completed job and a new job. Response time is the time a job spends in the box labeled "system."

## 1.3   LITTLE'S LAW

Consider a system with arrival and departure of jobs. Let $\lambda$ be the arrival rate (i.e., average inter-arrival time is $\frac{1}{\lambda}$), $T$ the average time a job spends in the system, and $\bar{n}$ the average number of jobs in the system. These three quantities, illustrated in Fig. 1.5, have a fundamental relationship:

**Little's Law:**      In steady state,      $\bar{n} = \lambda T.$

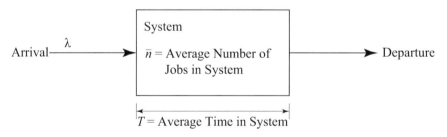

Figure 1.5: Little's Law relates these three quantities.

This is the most important result in performance analysis. It is very general and very useful. There are various intuitive arguments to support Little's Law, but a rigorous proof is nontrivial, since it must deal with delicate details like relating $\bar{n}$ as an average over time ($\bar{n} = \lim\limits_{T \to \infty} \frac{1}{T} \int_0^T n(t)dt$) to $\bar{n}$ as an average of a steady-state probability distribution ($\bar{n} = \sum\limits_{k=0}^{\infty} k\,\text{Prob}(n = k)$).

# 1.4    DISCUSSION OF PAPERS

We can now start to discuss the analytical models in 4 of the 40 papers. (We will use *italics* when referring to an equation, figure, table, section, etc. in these papers.) This discussion will be continued in later chapters as we cover further topics relevant to the models.

*StreamJoins* [28]

> The paper considers a system that includes a CPU, the memory, an index, and two streams. The input parameters include arrival rates $\lambda_a$ and $\lambda_b$, join operator service rate $\mu$, window sizes and memory size $M$. The main performance measures are the CPU time and throughput $r_o$.

> The paper focuses on two issues: (i) which join method (nested loop or hash) to use and (ii) how should CPU and memory be allocated?

> The model is an open system, but the paper does not say what the arrival process is. We will examine this point in the next chapter.

*MediaStreaming* [62]

> For this paper, the system has a set of streaming servers and peers connected by a network, and a collection of files for download. The analytical model does not include the directory server.

> The motivation is scalability: having multiple servers can relieve the bottleneck at a single central server, but that does not fundamentally address the necessity to increase capacity as more requests arrive. The idea here is to scale capacity by harnessing peer bandwidth, and this paper examines the scaling up rate.

> The table in *Sec. 3.2* of the paper lists the input parameters $L$, $N$, $n$, $\lambda$, $F$, $M$, $p_i$, $c_i$, and $b$, as well as the performance measures $k_0$ and $P(k)$.

> Given that $n$ is the number of peer classes and $b$ is the bandwidth required to stream a file, $N$ is a poor choice of **notation** for total server bandwidth. As in programming, the use of mnemonic variables can be a great help to a reader. In this case, using $b_{\text{servers}}$ for total server bandwidth and $b_{\text{file}}$ for the bandwidth to stream a file may be a better choice.

> The parameter $M$ for total peer population suggests that this is a closed model. In that case, the request rate should be a function of $M$, think time per peer, system response time, etc. (see Fig. 1.4). However, the model also has a parameter $\lambda$ for (total) request rate, which suggests that it is supposed to be an open model.

> We can see an application of Little's Law in *Eq. (3)*: Since request rate is $\lambda$ and each streaming session has length $L$, there are $\lambda L$ active streams at any time (if no request is rejected), and the necessary bandwidth to support them is $\lambda L b$.

> Note the reference to a measurement study in *Sec. 3.6* to justify using an exponential distribution to model peer lifespan.

*WebCrawler* [8]

The system here includes a set of pages out in the Web, the updates on these pages, the scheduled retrievals of these pages, and the local collection of the retrieved copies.

The issue is how to keep the copies as fresh (i.e., up-to-date) as possible by properly scheduling the retrievals. Aside from controlling the retrieval frequency, there is a choice of whether to download all pages in one batch periodically, or crawl selectively and incrementally.

The parameters are therefore the update rate $\lambda$ and crawl frequency $f$. *Fig. 2* presents measurements that show $\lambda$ differs for different domains, so one must also vary $f$ accordingly.

Since the issue is freshness, the authors first defined a metric for this. If $C = \{c_1, \ldots, c_N\}$ is the collection of local copies, the freshness of page $c_k$ at time $t$ is $F(c_k; t) = 1$ if $c_k$ is up-to-date at time $t$, and 0 otherwise. The freshness of $C$ is then $F(C; t) = \frac{1}{N} \sum_{k=1}^{N} F(c_k; t)$.

The authors use the exponential distribution to model the inter-update time. This distribution is easy to work with, and they were able to prove interesting results. For example, batch and incremental crawling yields the same average freshness if their average crawl speeds are the same; and if the update rate increases beyond some threshold, then the crawl speed should be *reduced*.

One might dismiss these results because they are based on the strong memoryless property, so it is crucial that the paper presents *Fig. 6* to show that the model agrees with the measurements.

Our intuition might suggest that the crawl rate $f_i$ for a page $e_i$ should depend on its change rate $\lambda_i$. For example, $f_i$ may be proportional to $\lambda_i$, so $f_i = c\lambda_i$, where $c = N / \left( I \sum_{k=1}^{N} \lambda_k \right)$, i.e., $f_i = \frac{\lambda_i}{\sum_{k=1}^{N} \lambda_k} \frac{N}{I}$, so the total crawl rate of $N$ pages over period $I$ is distributed among $e_i$ in proportion to their $\lambda_i$ value. Instead, Cho proved in his Ph.D. dissertation [7] that such proportionate crawling has poorer expected freshness than a uniform crawl rate of $f_i = 1/I$; this proof uses Jensen's inequality, and only requires convexity of $EF(e_i; t)$.

The underlying reason for this unintuitive result lies in $c = N/(I \sum_{k=1}^{N} \lambda_k)$, which constraints proportionate crawling to have $I \sum_{i}^{N} f_i = N$ crawls per period $I$, same as uniform. In fact, Cho used Lagrange multipliers to prove that, with this constraint, the crawl rate that maximizes average freshness is not uniform either, but follows the curve in *Fig. 9*. In particular, the figure shows that for $\lambda > \lambda_h$, the optimal crawl frequency $f_i$ *decreases* as change rate $\lambda_i$ increases. This is because, if a page is changing too fast, then it is not an optimal use of the budget ($N$ crawls over period $I$) to aggressively crawl that page.

These results (proportionate vs. uniform, decreasing $f_i$ for increasing $\lambda_i$) give an excellent illustration of how an analytical model can reveal unexpected insight and improve our intuition.

*SoftErrors* [40]

Soft errors are caused by radiation (e.g., cosmic rays). Their significance increases as features get smaller and voltage becomes lower in a chip. The *Soft Error Rate (SER)* can be estimated by summing the impact on various microarchitectural structures (reorder buffer, issue queue, etc.).

The impact on a structure $S$ is measured by an *Architectural Vulnerability Factor (AVF$_S$)*, so $SER = \sum_S AVF_S \times$ (intrinsic fault rate for $S$) (assuming the radiation faults occur randomly and uniformly over the chip). $AVF_S$ is the probability that a soft error in $S$ results in a program error. This probability depends on whether the error affects an ACE (*Architecturally Correct Execution*) bit.

The number of ACE bits in a structure changes over time as a workload executes. The AVF model in this paper therefore requires, as input, the microarchitectural specification and a profile of the workload, from which one extracts some input parameters ($\alpha$ and $\beta$ for critical path length, average instruction latency $\ell$, etc.).

Since the occupancy of ACE bits in $S$ is dynamic, the model divides time into intervals. Within each interval, occupancy is in steady state, and an interval starts and ends with some event (e.g., cache miss, misprediction) that changes the occupancy.

For occupancy at the reorder buffer (ROB), arrival rate is $I$ instructions per cycle (IPC), and each instruction stays for $\ell K(W)$ cycles, where $K(W)$ is the average critical path length for an instruction window of size $W$. By Little's Law, $W = I\ell K(W)$, and thus *Eq. (2)*. This gives the steady state, miss-free $O_{ideal}^{ROB}$ in *Eq. (3)*.

When the steady state is disrupted by an event, the paper models the change in occupancy by a straight line with slope $D$, the dispatch/retirement rate.

CHAPTER 2

# Single Queues

A typical computer system has queues everywhere. A queue imposes order on a set of tasks, making some wait for others. Such waiting is usually wasted time, so modeling system performance requires an understanding of queue behavior. This chapter introduces some elementary queueing theory.

## 2.1  APPLYING LITTLE'S LAW TO A 1-SERVER QUEUE

Consider a queue with one server and buffer space for holding jobs that are in service or waiting, as illustrated in Fig. 2.1. (Unless otherwise stated, a "server" refers to the server of a queue, rather than the server—web server, IO server, etc.—in some client-server architecture.)

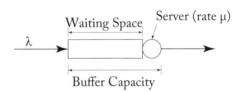

Figure 2.1: A simple queue with a single server. Queue length refers to all jobs at the queue, i.e., jobs waiting and in service. System time refers to total time spent at the queue, i.e., waiting time plus service time.

Let $\lambda$ and $\mu$ be the arrival and service rates, i.e., the average inter-arrival time is $1/\lambda$ and the average service time is $1/\mu$. Without specifying other details like the probability distributions for inter-arrival and service times and the queueing discipline, we can already apply Little's Law to derive some relationships (see Fig. 2.2).

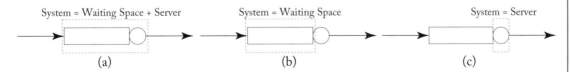

Figure 2.2: Little's Law is applicable to all three interpretations of "system."

(a) If we view the waiting space and server as the system, then Little's Law gives

$$\bar{n} = \lambda T, \tag{2.1}$$

where $\bar{n}$ is the average number of jobs at the queue (waiting and in service), and $T$ is the average time at the queue.

(b) If we view the waiting space alone as the system, then

$$L = \lambda W, \tag{2.2}$$

where $L$ is the average number of jobs waiting for service, and $W$ is the average waiting time.

(c) If we view the single server alone as the system, then the average number of jobs at the server is $1 \cdot \text{Prob}(1 \text{ job at the server}) + 0 \cdot \text{Prob}(\text{no job at the server}) = \lambda\left(\frac{1}{\mu}\right)$, so

$$\rho = \frac{\lambda}{\mu} \quad \text{where } \rho = \text{Prob}(1 \text{ job at the server}). \tag{2.3}$$

A server is busy if and only if it is serving a job, so $\rho$ is a measure of **server utilization**.

Since $T = W + \frac{1}{\mu}$ (waiting time plus service time), we have

$$\bar{n} = \lambda T = \lambda\left(W + \frac{1}{\mu}\right) = L + \rho. \tag{2.4}$$

As $\rho$ is a probability, $\rho = \frac{\lambda}{\mu}$ requires that $\lambda < \mu$, i.e., arrival rate is slower than service rate. This is intuitive; if $\lambda > \mu$, then jobs arrive faster than they can be served, the queue will grow indefinitely, there is no steady state, and averages have no meaning.

What if $\lambda = \mu$? The answer depends on whether the inter-arrival and service times are deterministic. Such details are specified by the notation we introduce next.

## 2.2   QUEUE SPECIFICATION

In the **Kendall Notation** $A/S/K/B/N/d$:

  $A$  is the arrival process (e.g., $M$ for memoryless, i.e., Poisson arrivals);

  $S$  is the service time distribution (e.g., $D$ for deterministic, $G$ for general);

  $K$  is the number of servers (possibly infinite);

  $B$  is the buffer capacity (including jobs in service; by default, $B$ is infinity);

  $N$  is the job population size (by default, $N$ is infinity);

*d* is the queueing discipline (by default, *d* is FCFS, i.e., first-come-first-served).

Another frequently used queueing discipline is **processor sharing**, where *n* jobs at a server will each get service rate $\frac{\mu}{n}$; this is an idealized model of round-robin scheduling.

Thus, $M/M/1$ denotes a queue with exponentially distributed inter-arrival and service times, 1 server, unlimited buffer capacity and population size, and FCFS scheduling. Another example: $M/D/3$ denotes a queue with Poisson arrivals, deterministic service times, and 3 servers, as illustrated in Fig. 2.3. If the service rate is $\mu$ for each server, then the service rate for the queue would be $\mu$ if it has only 1 job, $2\mu$ if it there are 2 jobs, and $3\mu$ if there are 3 or more.

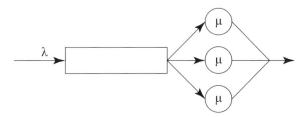

Figure 2.3: $M/D/3$ has Poisson arrivals, deterministic service times, three servers, and infinite buffer capacity.

Note the implicit assumption here that a job will never wait if there is an idle server. This is an example of a **work conserving** queue. More sophisticated queueing disciplines may violate this assumption; for example, to avoid a situation where long jobs hog all servers, one server may be dedicated to short jobs and therefore allowed to idle if all jobs in the queue are long.

Note also that, for a queue with $K$ servers, steady state is possible for $\lambda > \mu$—as long as $\lambda < K\mu$, so that utilization per server $\frac{\lambda}{K\mu}$ is well defined.

Although $M/M/\infty$ denotes a queue with infinitely many servers, it is an abstraction for a queue where jobs do not have to wait for each other: each arriving job stays for an exponentially distributed duration, then leaves. An $M/M/\infty$ queue is therefore sometimes called a **delay center**.

## 2.3    POLLACZEK–KHINCHIN FORMULA

Beyond the basic queues represented by the Kendall notation, there are a myriad other variations, with a huge literature to match. It is therefore surprising that there is in fact a general formula for $M/G/1$.

**Pollaczek–Khinchin Mean-Value Formula**

Let $C$ be the coefficient of variation for the service time of $M/G/1$ (Definition 0.4).
Then

$$\bar{n} = \rho + \rho^2 \frac{1 + C^2}{2(1 - \rho)} \quad \text{where } \rho \text{ is the server utilization.}$$

Recall from Eq. (2.3) that $\rho = \frac{\lambda}{\mu}$ where $\lambda$ and $\mu$ are arrival and service rates. Although the Pollaczek–Khinchin formula is stated here for the default queueing discipline (FCFS), it has been generalized to other disciplines.

The Pollaczek-Khinchin formula shows that average queue size $\bar{n}$ is near 0 for small $\rho$, but grows rapidly as $\rho$ approaches 1. This is illustrated in Fig. 2.4.

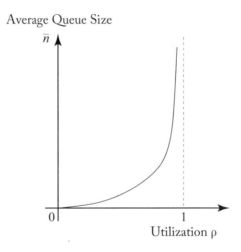

Figure 2.4: $M/G/1$: Average queue size is near 0 for small utilization, but grows rapidly as $\rho$ approaches 1.

By Little's Law, the average system time $T = \bar{n}/\lambda$, so the Pollaczek–Khinchin formula gives

$$T = \frac{1}{\mu} + \frac{1}{\mu} \rho \frac{1 + C^2}{2(1 - \rho)}.$$

We therefore get the behavior in Fig. 2.5 as arrival rate increases: $T$ is basically service time $\frac{1}{\mu}$ for small $\lambda$, but goes to infinity as $\lambda \to \mu$ (i.e., $\rho \to 1$). This shows graphically how $M/G/1$ becomes unstable as $\lambda$ approaches $\mu$, so steady state is impossible for $\lambda = \mu$.

Why does an $M/G/1$ queue blow up this way for $\lambda < \mu$ (after all, the server still has some idle time since $\rho \neq 0$)? This behavior is caused by the randomness in the service and arrival

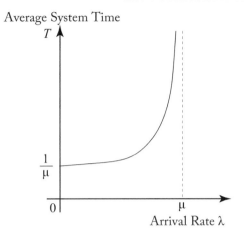

Figure 2.5: $M/G/1$: Average system time is near service time $\frac{1}{\mu}$ for small arrival rate, but grows rapidly as $\lambda$ approaches service rate $\mu$.

times. For example, if service times are random, then as $\rho$ increases (the server becomes busier), it becomes more likely that the queue will contain a long job, whose service time will be added to every job that has to wait for it.

**Example:** $M/M/1$
By Lemma 0.2, an exponential distribution has $C = 1$. It follows that, for $M/M/1$,

$$\bar{n} = \rho + \rho^2 \frac{1+1}{2(1-\rho)} = \frac{\rho}{1-\rho}. \tag{2.5}$$

From Eq. (2.1),

$$T = \frac{\bar{n}}{\lambda} = \frac{1}{\mu(1-\rho)}. \tag{2.6}$$

From Eq. (2.4),

$$L = \bar{n} - \rho = \frac{\rho^2}{1-\rho}. \tag{2.7}$$

From Eq. (2.2),

$$W = \frac{L}{\lambda} = \frac{\rho}{1-\rho}\frac{1}{\mu}. \tag{2.8}$$

If service time is deterministic (i.e., $M/D/1$), then $C = 0$, so

$$\bar{n} = \rho + \rho^2 \frac{1+0}{2(1-\rho)} = \left(1 - \frac{\rho}{2}\right)\frac{\rho}{1-\rho} \quad \text{for } M/D/1. \tag{2.9}$$

Compare this to Eq. (2.5), and we see that (for the same server utilization $\rho$) $M/D/1$ has smaller queue size $\bar{n}$. This illustrates the impact of randomness in service times on performance.

In fact, for $D/D/1$, it is easy to see that

$$T = \frac{1}{\mu}, \quad \bar{n} = \frac{\lambda}{\mu}, \quad L = 0 \quad \text{and} \quad W = 0 \quad \text{for any } \lambda \leq \mu.$$

Here, the existence of a steady state for $\lambda = \mu$ illustrates the impact of determinism in arrival times.

## 2.4   DISCUSSION OF PAPERS

We continue the discussion of *StreamJoins* [28] and introduce *SleepingDisks* [68], *StorageAvailability* [18], *DatacenterAMP* [22], and *GPU* [26].

*StreamJoins* [**28**]

For limited compute power (*Sec. 5.2.1*), the streams are sampled, resulting in effective arrival rates of $\lambda_{a'}$ and $\lambda_{b'}$. Their inter-arrival times are random variables (even if $\lambda_a$ and $\lambda_b$ are deterministic), since $\lambda_{a'}$ and $\lambda_{b'}$ are obtained by random sampling. If the streams are Poisson and the tuples are sampled independently with a fixed probability, then the Split Property of Theorem 0.16 says that the sampled streams are also Poisson.

However, it appears from $\lambda_{a'} + \lambda_{b'} = \mu$ in *Eq. (13)* that the authors assume the stream arrivals and service times are deterministic; otherwise, the average number of tuples in memory can grow without bound, like in Fig. 2.4, so no amount of memory will be enough for the case of "unlimited memory." From Eq. (2.9), we see that even for $M/D/1$, $\lim_{\rho \to 1} \bar{n} = \infty$.

Note that data streams like stock price movements, news stories, and network measurements are not likely to have deterministic inter-arrival times.

For unlimited memory, one could use Little's Law to calculate the memory requirement from $\lambda_{a'}$, $\lambda_{b'}$ and the time a tuple spends in memory. However, this time depends on the service time distribution (not just the average $\mu$—compare Eq. (2.5) and Eq. (2.9)). The tuple processing time includes probing, insertion, and invalidation—which vary from tuple to tuple—so the service time is not deterministic.

In the case of limited memory and unlimited computing resources (*Sec. 5.2.2*), if the streams are not deterministic, then a burst of arrivals can cause the buffer to overflow and tuples may be lost, so the arrival rates at the join may no longer be $\lambda_a$ and $\lambda_b$. Similarly, if the computing resources are also limited (*Sec. 5.2.3*), then the sampled streams may not have the intended $\lambda_{a'}$ and $\lambda_{b'}$.

*SleepingDisks* [68]

Fig. 6 shows the system for this paper. The input parameters include arrival rate $\alpha_i$, service time as specified by $Exp(t_{ij})$ and $Var(t_{ij})$, transition time $T_i$, power $P_{ij}$, $P'_{ij}$, and $P''_{ij}$. The main performance measures are response time (*Fig. 7*), transaction rate (*Fig. 12a*), and energy per transaction (*Fig. 12b*).

The main issues are: (i) the tradeoff between performance and energy and (ii) adaptive disk layout in response to dynamic workload changes.

Note the use of Little's Law (Eq. (2.3)) and the Pollaczek–Khinchin formula in deriving $R''_{ij}$. The expressions for $R'_{ij}$ and $R'''_{ij}$ use the residual life introduced in the next chapter.

*StorageAvailability* [18]

Fig. 1 shows the StarFish system, consisting of a host element HE, storage elements SEs, and the network that connects them. The data replication through multiple SEs is for fault tolerance, and the varied location distances from the HE seek to balance high resilience and low latency.

The model analyses the reliability and availability issues. When a model is meant to study **rare events**, simulation models can take too long to generate those events, so there is little choice but to use an analytical model.

The input parameters are the number of SEs $N$, the write quorum $Q$, the failure rate $\lambda$ and recovery rate $\mu$. Note that unlike an open queue (e.g., $M/G/1$) it is possible that $\lambda \geq \mu$, so $\rho \geq 1$: that just means that the machine is more often down than up. The performance metrics are **availability** $A(Q, N)$, and reliability as measured by the probability that there is no data loss.

As *Table 1* shows, availability is measured in *nines*, e.g., six nines is 99.9999% availability. The precision suggested by such multiple significant figures may look impressive, but one must not forget the underlying assumptions: in this model, for example, the six nines are calculated with exponential distributions. An earthquake can also break the links to two or more SEs (StarFish considers network disconnection to be a site failure), so failures may not be independent.

*DatacenterAMP* [22]

Power consumption and latency bounds are two fundamental issues for current datacenters. This paper studies the use of asymmetric multicore processors (AMP) to address these issues (*Fig. 1*): a smaller, less power-intensive core is used if the service level agreement (SLA) can still be met (Energy Scaling); and core asymmetry can be adjusted to bring task execution within SLA (Parallel Speedup).

There were previous analyses of AMP performance [25]. This paper is different in that it considers how AMP affects a stream of tasks (i.e., not just one task). Specifically, tasks have nondeterministic execution time, and any dynamic adjustment of core asymmetry would add to the nondeterminism. These introduce a queueing effect that can hurt SLA-compliance. In contrast, *StreamJoins* [28] does not model the queueing effect.

The system here is a simple queue with an AMP as server, and a task stream. The key model parameters are the latency SLA $T_{SLA}$, total area budget $n$, core area $r$, and task fraction $f$ that can be parallelized. Naturally, the performance metrics are speedup and power savings.

The latency model is a simple $M/M/1$ queue. By Eq. (2.6), and since $\rho = \lambda/\mu$, the expected latency is $\frac{1}{\mu(1-\rho)} = \frac{1}{\mu-\lambda}$, as per *Eq. (1)*. The power proportional model in *Eq. (5)* comes from

$$(P_{CPU} + P_{other})\text{Prob(busy CPU)} + P_{idle}\text{Prob(idle CPU)},$$

where Prob(busy CPU)$= \rho = \lambda/\mu$. They further model $P_{CPU}$ as proportional to area, and $P_{other}$ as proportional to $\rho$.

The paper adopts Hill and Marty's modification of Amdahl's Law [25] to model the speedup. With neat closed-form expressions like *Eq. (3)* and *Eq. (4)*, the authors could have derived how the optimal large core size $r$ depends on the parallelizable fraction $f$, and thus gain further insight into *Fig. 4*; e.g., why the best speedup ratio is not monotonic in $r$—see how the curves for $r = 32$ and $r = 64$ intersect.

Similarly, the closed-form *Eq. (5)* can be analyzed to show how the maximum possible power savings varies with SLA for given $f$, $n$, and $r$. Such an analysis will also reveal the relative impact of $P_{CPU}$, $P_{other}$ and $P_{idle}$. This impact is hidden once they **normalize** the power to $P_{CPU} = 1$ and use the magic value $P_{idle} = 0.1$.

Using $M/M/1$ to model the CPU queue is very simplistic, possibly casting doubt on the validity of the conclusions. It is therefore crucial that the authors test the model experimentally. The agreement in *Fig. 2* between values calculated with the model and measured in the experiments is impressive, and strongly supports their results.

## GPU [26]

A GPU runs multiple threads simultaneously. These threads are grouped into *warps*. For this paper, the warps run one at a time, and the threads in a warp execute the same instruction together, in program order. If some thread in a warp stalls because of memory divergence, execution can pass to another warp. The warps compete for memory, in the form of MSHRs (Miss Status Handling Registers) and the DRAM bus. This resource contention degrades performance, thus balancing the performance improvement from multithreading (*Eq. (3)*).

The system in the model has warps and queues for MSHR and DRAM. The input to the model includes warp profiles, number of MSHRs, miss latency, cache line size, DRAM bandwidth, etc.

Like *SoftErrors* [40], the model discretizes time by using cache miss events to divide time into intervals. However, the interval analysis here is not aimed at decomposing a dynamic state into intervals of steady state; rather, it facilitates the multithreading analysis for warp scheduling (*Fig. 8*).

A warp is hence modeled with an interval profile (*Eq. (2)*) that is highly abbreviated. Even so, two threads with the same instruction sequence may branch differently, so this control divergence generates multiple warp profiles. GPUMech handles this diversity by further condensing each interval profile into a 2-dimensional (2D) vector (*Eq. (6)*), and applies 2-means clustering to select a representative warp. It illustrates the use of *data mining* in performance modeling.

There are three parts to the model for analyzing a representative warp. The first part models warp scheduling (Round Robin or Greedy-Then-Oldest) by using the concept of an *issue probability* (that is estimated with a uniform distribution in *Eq. (9)*). The second part uses *Eq. (19)* to estimate the MSHR queueing delay, assuming the memory instructions are issued by all warps in the same cycle. It is not clear why the authors did not relax this assumption by using a multiserver queue.

The third part uses an $M/D/1$ queue to model the DRAM delay—the expression $\frac{\lambda s^2}{2(1-\rho)}$ in *Eq. (21)* can be derived from Eqs. (2.2), (2.4), and (2.9). *Fig. 15* shows that, without modeling this delay, GPUMech's accuracy will be severely affected as DRAM bandwidth decreases.

The model's accuracy also depends on the workload. *Fig. 16* shows a significant gap in CPI between simulated measurements and model calculation for `kmeans_invert_mapping`. For this workload, the CPIstack breakdown indicates DRAM queueing delays that are much greater than the DRAM bus transmission time. This suggests a high utilization (Fig. 2.5) that pushes the model to its limits.

# CHAPTER 3

# Open Systems

In performance analysis, we are often interested in not just average values, but probabilities too. This chapter shows how such details can be obtained for queues in an open system.

We first introduce the concept of residual life, which is often used implicitly in performance models.

## 3.1 RESIDUAL LIFE

Consider a system in which events are occurring at times $T_1 < T_2 < \cdots$. An observer arrives at a random point in time $t$ (i.e., $t, T_1, T_2, \ldots$ are independent) and observes the next event at $T_n$; $T_n - t$ is called a **residual life**. This is illustrated in Fig. 3.1.

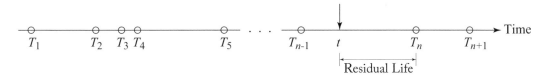

Figure 3.1: Events occur at times $T_1, T_2, \ldots$, and an arrival at time $t$ observes the next event at $T_n$. $T_n - t$ is a residual life.

What is $E(T_n - t)$? Assume exponential inter-event times, $T_{i+1} - T_i \sim \text{Exponential}(\mu)$. One might argue that, by the memoryless property,

$$E(T_n - t) = \frac{1}{\mu}. \tag{3.1}$$

On the other hand, since $t$ is arbitrary, we expect

$$E(T_n - t) = \frac{1}{2} E(T_{i+1} - T_i) = \frac{1}{2\mu}, \tag{3.2}$$

which contradicts Eq. (3.1). From renewal theory [30], if $X$ is the inter-event time, then

$$\text{expected residual life} = E(T_n - t) = \frac{EX^2}{2EX} = \frac{EX}{2} + \frac{VarX}{2EX}, \tag{3.3}$$

since $EX^2 = VarX + (EX)^2$. By Lemma 0.2, we get

$$E(T_n - t) = \frac{1}{2\mu} + \frac{\left(\frac{1}{\mu}\right)^2}{2\left(\frac{1}{\mu}\right)} = \frac{1}{\mu},$$

so Eq. (3.1) is correct. To resolve the contradiction, note that the derivation in Eq. (3.2) should be

$$E(T_n - t) = \frac{1}{2} E(T_n - T_{n-1}), \quad \text{so } E(T_n - T_{n-1}) = 2E(T_n - t) = \frac{2}{\mu},$$

i.e., $t$ is more likely to fall within a larger interval $T_{i+1} - T_i$.

Note from Eq. (3.3) that, if $VarX \approx 0$, then the expected residual life is approximately $EX/2$, which is intuitive. In particular, if the inter-event time is deterministic ($VarX = 0$) and $T_{i+1} - T_i = \Delta$ (constant), then the expected residual life is $\frac{1}{2}\Delta$.

However, if $T_{i+1} - T_i$ is uniformly distributed between 0 and $2\Delta$ (so $E(T_{i+1} - T_i) = \Delta$), then

$$EX^2 = \int_0^{2\Delta} \frac{x^2}{2\Delta} dx = \frac{4}{3}\Delta^2,$$

so the expected residual life is $\frac{\frac{4}{3}\Delta^2}{2\Delta} = \frac{2}{3}\Delta$. We see here the difference between deterministic and randomly distributed inter-event times.

We will revisit residual life when we consider closed systems.

## 3.2    BIRTH-DEATH PROCESS

We now introduce the technique of birth-death processes by solving the $M/M/1$ queue. Unless otherwise stated, "queue" refers to the server, the buffer space and all jobs (waiting and in service).

Let $n$ be the number of jobs at the queue. Intuitively, $n$ suffices to specify the state of the system, since the arrivals and service are independent and memoryless, and scheduling is FCFS: If service were not memoryless, the state may need to record how much service a job has already received; if scheduling were not FCFS, the state may need to reorder the queue when a new job arrives.

Hence, the **state space** is $\{0, 1, 2, \ldots\}$. The system changes state from $n$ to $n + 1$ when a job arrives, and this transition occurs with arrival rate $\lambda$; similarly, the queue changes state from $n$ to $n - 1$ when a job completes, and the transition occurs with service rate $\mu$. This is illustrated by the **state transition diagram** in Fig. 3.2; it is an example of a **continuous-time Markov Chain**.

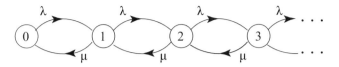

Figure 3.2: State transition diagram for $M/M/1$.

For inter-arrival times that are exponentially distributed, jobs can only arrive one at a time. Similarly, with a single server, FCFS and exponentially distributed service times, no two jobs can complete simultaneously. Thus, there are no transitions between any $n$ and $n'$ for $|n - n'| \geq 2$. Such a Markov chain is also called a **birth-death process**.

Let $p_k = \text{Prob}(n = k)$. To solve for $p_k$, we can draw a boundary between states 0 and 1, as illustrated in Fig. 3.3.

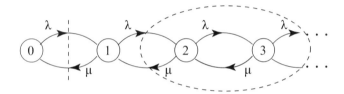

Figure 3.3: Expected net flow across a boundary is 0.

In steady state, we expect the net flow across this boundary to be 0, i.e., $p_0 \lambda - p_1 \mu = 0$. By similarly drawing boundaries between any states $n$ and $n + 1$, we thus get

$$p_k \lambda = p_{k+1} \mu \qquad \text{for } k = 0, 1, 2, \ldots. \tag{3.4}$$

These are examples of **local balance equations**. One could extend this idea to derive equations by drawing any boundary to enclose a set of states. For example, if the boundary encloses states 2 and 3 in Fig. 3.3, we get

$$p_1 \lambda + p_4 \mu = p_2 \mu + p_3 \lambda. \tag{3.5}$$

This is an example of a **global balance equation**. However, Eq. (3.5) is a sum of two local balance equations (for $k = 1$ and $k = 3$), so it provides no new information.

To solve the local balance equations, let $\rho = \frac{\lambda}{\mu}$. Then Eq. (3.4) gives

$$p_{k+1} = \rho p_k \qquad \text{for } k = 0, 1, 2, \ldots,$$

$$\text{so} \quad p_k = \rho^k p_0 \qquad \text{for } k = 0, 1, 2, \ldots.$$

The state probabilities must sum to 1, i.e., $\sum_{k=0}^{\infty} \rho^k p_0 = 1$,

$$\text{so} \quad p_0 = \frac{1}{\sum_{k=0}^{\infty} \rho^k} = 1 - \rho \tag{3.6}$$

$$\text{and} \quad p_k = \rho^k (1 - \rho) \qquad \text{for } k = 1, 2, 3, \ldots . \tag{3.7}$$

Note that, as there is only one server, Eq. (3.6) gives Prob(1 job at the server)$= 1 - p_0 = \rho$, in agreement with Eq. (2.3). Moreover,

$$\bar{n} = \sum_{k=0}^{\infty} k p_k = \sum_{k=1}^{\infty} k \rho^k (1 - \rho) = \rho \sum_{k=1}^{\infty} k \rho^{k-1} (1 - \rho) = \frac{\rho}{1 - \rho} \quad \text{by Lemma } 0.7$$

since $\rho^{k-1}(1 - \rho)$ is a geometric distribution. We have thus derived Eq. (2.5) for $M/M/1$ without using the Pollaczek–Khinchin formula.

**Example:** $M/M/3/5$

If we limit the buffer capacity of an $M/M/3$ queue to five jobs, then we get the $M/M/3/5$ queue in Fig. 3.4a; Fig. 3.4b shows the corresponding transition diagram.

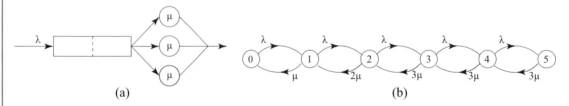

(a)                 (b)

Figure 3.4: The $M/M/3/5$ queue in (a) has the transition diagram in (b).

**Example:** $M/M/K/K$

For a queue with finite buffer capacity, it is possible that an arriving job finds the buffer full; a common assumption is that such a job is lost or rejected. A birth-death process can be used to calculate the probability of such an event. For example, one can use the transition diagram for $M/M/K/K$ (Fig. 3.5) to derive

$$\text{Prob(job is lost)} = \text{Prob}(n = K) = \frac{\frac{\rho^K}{K!}}{\sum_{r=0}^{K} \frac{\rho^r}{r!}}. \tag{3.8}$$

This is called **Erlang's B** formula for **blocking probability**. It in fact holds for $M/G/K/K$ [23].

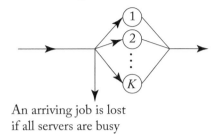

An arriving job is lost
if all servers are busy

Figure 3.5: Jobs that arrive at a finite-buffer queue will be lost if the buffer is full.

## 3.3 OPEN QUEUEING NETWORKS: JACKSON NETWORKS

Consider an open queueing network, like in Fig. 1.1. Assume each queue has one or more servers with exponentially distributed service time, and Poisson arrival of jobs to the network. (By Theorem 0.16, we may assume they form a single stream that possibly splits into Poisson arrivals at multiple queues.) This is an example of a **Jackson network** [27].

If the network has $Q$ queues and if $n_i$ is the number of jobs at queue $i$, then a Jackson network in steady state has the surprising property that

$$\text{Prob}(n_1 = k_1, \ldots, n_Q = k_Q) = \text{Prob}(n_1 = k_1) \cdots \text{Prob}(n_Q = k_Q). \tag{3.9}$$

In other words, despite queues feeding into one another, the queues behave like they are independent, in the sense that the joint probability in Eq. (3.9) is just a product of marginal probabilities. For this reason, such a network is called **product-form**, or **separable**.

In fact, each queue behaves like an independent $M/M/K$ or $M/M/\infty$ queue, so we can use a Markov chain to derive an expression for each $\text{Prob}(n_i = k_i)$ separately and substitute it into Eq. (3.9). However, this does not mean that each queue has Poisson arrivals; if there are loops, like in Fig. 1.1, then the arrival process at some queues may not be Poisson. This is why the product form in Eq. (3.9) is surprising.

**Example:**
Consider the open network in Fig. 3.6 with single-server queues $A$, $B$, and $C$. Suppose arrival rate to the system is $\lambda$, and a job departing $A$ chooses to visit $A$, $B$, or $C$ with **branching probability** $p_A$, $p_B$, and $p_C$, respectively.

Let $p_{\text{done}} = 1 - (p_A + p_B + p_C)$. Then the number of visits to $A$ has distribution Geometric($p_{\text{done}}$). Let $V_A$, $V_B$, and $V_C$ be the expected number of visits to $A$, $B$, and $C$, respec-

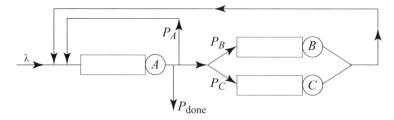

Figure 3.6: Example of an open system for application of Jackson's Theorem.

tively. Then

$$V_A = \frac{1}{p_{\text{done}}} \quad \text{by Lemma } 0.7,$$

$$V_B = p_B V_A = \frac{p_B}{p_{\text{done}}} \quad \text{and} \quad V_C = \frac{p_C}{p_{\text{done}}} .$$

Let $\lambda_A$, $\lambda_B$, and $\lambda_C$ be the job arrival rates at $A$, $B$, and $C$, respectively. Then

$$\lambda_A = V_A \lambda = \frac{\lambda}{1 - p_A - p_B - p_C} , \quad \lambda_B = V_B \lambda = \frac{p_B \lambda}{1 - p_A - p_B - p_C} ,$$

$$\lambda_C = \frac{p_C \lambda}{1 - p_A - p_B - p_C} .$$

Finally, let $n_A$, $n_B$, and $n_C$ be the number of jobs at $A$, $B$, and $C$, respectively. Since each queue has one server, we can now use $\lambda_A$ and the service rate $\mu_A$ to determine $\text{Prob}(n_A = h)$ with Eq. (3.7), and similarly for $\text{Prob}(n_B = i)$ and $\text{Prob}(n_C = j)$. Application of Jackson's Theorem and Eq. (3.9) then gives $\text{Prob}(n_A = h, n_B = i, n_C = j)$.

## 3.4 DISCUSSION OF PAPERS

In addition to further comments on previous papers, we introduce four more: *GPRS* [42], *TransactionalMemory* [24], *RouterBuffer* [1], and *ProactiveReplication* [13].

*SleepingDisks* [68]

The approximation for $R'_{ij}$ uses $T_i/2$ for the transition time, which agrees with the residual life of a deterministic inter-event time seen by a Poisson arrival.

In the approximation for $R'''_{ij}$, the first two terms for delay correspond to the Pollaczek–Khinchin formula; they should be divided by arrival rate $\alpha_i$ to give the waiting time (using

Little's Law) in the queue of foreground requests. The third term $\frac{E(t_{ij})}{2}$ is the expected waiting time for a background request; since $t_{ij}$ is not deterministic, this term should instead use the residual life expression $\frac{E(t_{ij})}{2} + \frac{Var(t_{ij})}{2E(t_{ij})}$.

*MediaStreaming* [62]

The paper does not analyze the probability that a request is rejected (when all servers and peers are busy). An obvious possibility is to use Erlang's formula. When a request is rejected, the peer is likely to repeat the request; the impact of such re-requests is not considered in the model.

*StorageAvailability* [18]

The classical machine repairman model is shown in Fig. 3.7a, where there is just one repairman, so there is a queue for concurrent failures. In the StarFish context, the SEs are repaired independently, so there are dedicated repairmen, like in Fig. 3.7b. The crucial difference between these two is that there is no queueing in Fig. 3.7b.

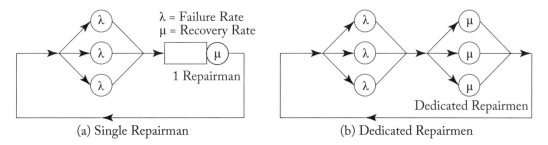

    (a) Single Repairman            (b) Dedicated Repairmen

Figure 3.7: Machine repairman models.

*GPU* [26]

For the bandwidth contention model, GPUMech bounds the delay computed with the Pollaczek-Khinchin formula for an $M/D/1$ queue, using the expression ($s \times$ #core_reqs$_i \times$ #cores)$/2$. This puts the representative warp in the middle ($1/2$) of the worst case, similar to the argument for the residual life for deterministic events.

*GPRS* [42]

Although the GPRS infrastructure has multiple overlapping wireless cells managed by base stations wired to a backbone of routers, this paper focuses on the timeslot allocation for downlink traffic to mobile nodes in one bottleneck cell.

The input parameters are $T$ (fixed number of TDMA timeslots), $N$ (number of mobile nodes in the cell), $d$ (number of slots that can be used simultaneously for downlink traffic), $t_B$ (time duration of one radio block), $x_B$ (number of bytes that can be transferred per block), $x_{on}$ (average number of bytes in a download), and $t_{off}$ (average duration of OFF period). The main performance measures are utilization $\bar{u}$, throughput $\bar{x}$ and blocking probability.

The main issue lies in determining the maximum number of active nodes in a cell ($n_{\max}$) and in $P$ cells ($M_{\max}$).

Note the argument in *Sec. 2.2* for adopting memoryless distributions for the ON period and OFF period: If the performance metric of interest (e.g., throughput or average number of active nodes) is sensitive only to the average values of the parameters (e.g., $x_{on}$ and $t_{off}$), then one might as well assume the distribution is memoryless, whose strong properties will help simplify the analysis.

The OFF period is the inter-arrival time, with an average of $\frac{1}{\lambda} = \frac{t_{off}}{t_B}$ blocks. The ON period is the download time, with an average of $\frac{1}{\mu} = \frac{x_{on}}{x_B}$ blocks. The ratio $x$ in *Eq. (6)* is thus $\frac{\lambda}{\mu}$.

The transition between two states is a probability $p_{ij}$ (not a rate, like jobs per second). Such a model is a **discrete-time Markov chain**, where time is discretized into timeslots, so $p_{ij}$ is the probability that the system goes from state $i$ at time $t$ to state $j$ at time $t + 1$. The balance equations are usually written as a vector equation with a transition matrix, and can be solved with software for linear algebra. There is a memoryless assumption here that this probability does not depend on states prior to time $t$.

*TransactionalMemory* [24]

Single-processor techniques for coordinating access to shared memory (e.g., using semaphores) do not scale well as the number of cores increases. A currently favored alternative is transactional memory, and this paper studies the performance impact when it is implemented in software.

The main parameters for the model are the number of data items $L$, the number of concurrent transactions $N$, the number of lock requests $k$—which are assumed to be uniformly distributed over the $L$ items—and the probability $\ell_w$ that a lock request is for write access.

The performance measures are number of locks held $E(Q)$, number of restarts $E(R)$ and number of lock requests $E(S)$. Here, $E(S)$ is a measure of response time. Since the authors use a closed model, one can deduce the throughput from $E(S)$ and $N$ (using Little's Law).

For a closed model, one can think of a committed transaction as releasing its locks and starting again as another transaction, so there is a transition from state $k + 1$ in *Fig. 1* back to state 0. This means state $k + 1$ is no longer an absorbing state, so there is a steady state.

This modified Markov chain can then be viewed differently, not as transitions among states of one tagged transaction, but as a "flow diagram" for the entire system, so node $j$ represents all transactions holding $j$ locks. Viewed this way, the equations can then be derived with just Little's Law (e.g., compare *Eq. (10)* in the paper and *Eq. (2.7)* in [58]). This avoids the need to appeal to the theory for finite Markov chains in the derivation of *Eq. (6)*, and the awkward mismatch between using a discrete time model when inter-request time in fact has a continuous distribution.

*RouterBuffer* [1]

Since this paper is about sizing router buffers for TCP, the system is, strictly speaking, TCP plus the Internet. However, as TCP behavior is largely determined by the bottleneck link, the paper zooms in on a single router, with buffers for its input and output links (*Fig. 2*). The issue is: What should the size of the buffer be for the bottleneck, if the link is not to become idle.

The parameters are the buffer size $B$, bottleneck bandwidth $C$, the round trip time $\overline{RTT}$, and the number of flows $n$.

The key result is that a link requires a buffer size of no more than $B = \overline{RTT} \times C / \sqrt{n}$, where $n$ is the number of long flows. This appears unintuitive: Shouldn't the buffer size increase with $n$, instead of decrease? However, the intuition becomes clear once we remember that, if a buffer is sized to prevent underflow, then since underflow is less likely for bigger $n$, buffer size can be smaller. Note the application of Little's Law in *Eq. (1)*—the number of bytes in flight is estimated as $2T_p C$, where $T_p$ is the one-way propagation delay.

*Sec. 3* examines what the buffer size should be to prevent underflow, if flows are long enough that TCP is in congestion avoidance. The goal in *Sec. 4*, however, is to avoid dropping packets—i.e., buffer overflow—for short flows that remain in TCP slow start.

The authors use $M/G/1$ to model the router for short flows. Although flows may arrive as a Poisson process (as empirically observed), the packets within a flow are synchronized by $RTT$ and have correlated sizes; therefore, the burst arrivals are not Poisson, and their service times are not independent. However, these correlations are weakened when bursts from a large number of flows are interleaved.

By default, an $M/G/1$ queue has an infinite buffer. The paper uses $\text{Prob}(Q \geq B)$ from the queue length distribution to approximate the probability of an overflow of the router's finite buffer.

*ProactiveReplication* [13]

From the modeling point of view, the system in this paper is a set of files stored in a peer-to-peer system, where each peer can disconnect permanently, or leave temporarily and return. Each file $F$ is broken into fragments, from which $F$ can be reconstructed. This storage system aims to ensure each file has $n'$ fragments online at all times.

Each peer holds at most 1 fragment of $F$. Replicating a fragment held by a peer that disconnects will generate network traffic. To smoothen the replication traffic caused by a spike in disconnections, this paper proposes a control mechanism (*Fig. 1*) to anticipate disconnections and proactively replicate fragments, at the cost of possibly having more fragments than necessary (*Fig. 13*).

Aside from the target $n'$, the model takes as input a log of peer disconnections and reconnections. Time is divided into intervals of varying length $\Delta T$. *Eq. (4)* calculates for $\Delta T$ the number $\hat{n}$ of available fragments, and *Eq. (5)* calculates the rates $\hat{\gamma}_1$ and $\hat{\mu}$. There is no way of telling if a disconnection is going to be permanent, but the authors get around this by using *Eq. (6)* to calculate the abandon probability $\hat{P}$ using the repair rate in $\Delta T$. The parameter $D$ is used to resize $\Delta T$, using *Eq. (8)*.

The queueing network is actually not used in these calculations, which just need Little's Law (*Eq. (2), Eq. (5), Eq. (8)*), assuming the system is in steady state within each $\Delta T$. If the repairs are Poisson arrivals (rate $R$), and connection/disconnection times are exponentially distributed, then *Fig. 3* is an open Jackson network, so it can be decomposed into two $M/M/\infty$ queues; for large $\bar{n}$, the number of jobs in an $M/M/\infty$ queue is approximately Poisson distributed (*Eq. (3)*). This distribution (or its Normal approximation) can be used to set Service Level Objectives, like $\text{Prob}(n < n^*) < \epsilon$ for some target $n^*$ and $\epsilon$.

# CHAPTER 4

# Markov Chains

A birth-death process is an example of a 1-dimensional (1D) Markov Chain. Markov chains are applicable far beyond queueing analysis: They have a flexibility that makes them easy to use when analyzing the performance of disparate systems. This chapter looks at examples of two multidimensional Markov chains.

## 4.1 MARKOV CHAIN FOR A CLOSED NETWORK

Consider a network of three queues, similar to the one in Fig. 1.1 but closed, as shown in Fig. 4.1.

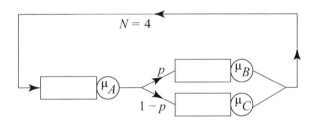

Figure 4.1: A closed network with three queues and four jobs.

Let $n_A$, $n_B$, and $n_C$ be the number of jobs at queues $A$, $B$, and $C$, respectively, and $p$ the branching probability that a job departing from queue $A$ chooses to go to queue $B$. Suppose there are 4 jobs, so $n_A + n_B + n_C = 4$. If service times are memoryless, the state space is $\{\langle n_A, n_B, n_C \rangle \mid n_A + n_B + n_C = 4\}$; equivalently, we can use $\{n_B n_C \mid 0 \le n_B + n_C \le 4\}$. As shown in Fig. 4.2, the corresponding Markov chain is 2D. Note that (like in the case of birth-death processes) transitions between states 12 and 21, or between 12 and 00, say, are impossible because they require simultaneous events.

This Markov chain has 15 variables, namely $p_{ij} = \text{Prob}(n_B = i \text{ and } n_C = j)$, where $0 \le i + j \le 4$. Aside from $\sum_{i,j} p_{ij} = 1$, we can derive 14 independent equations by balancing flows across various boundaries, like in Fig. 4.3. For example, the boundary $\beta$ between state 30 and

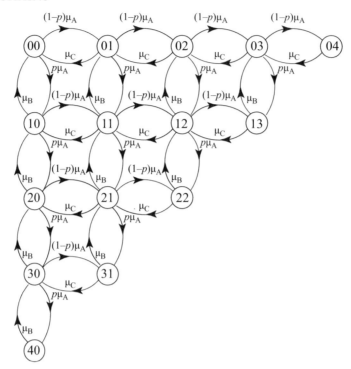

Figure 4.2: The Markov chain for the queueing network in Fig. 4.1 is 2D.

states 20 and 31 in Fig. 4.3 gives

$$p_{30}\mu_B + p_{30}(1-p)\mu_A = p_{31}\mu_C + p_{20}p\mu_A. \tag{4.1}$$

However, if we look at Fig. 4.1 itself, we see that the steady-state job flow into and out of queue $B$ must balance, giving us

$$p_{30}\mu_B = p_{20}p\mu_A. \tag{4.2}$$

Similarly, the flow balance for queue $C$ gives

$$p_{31}\mu_C = p_{30}(1-p)\mu_A. \tag{4.3}$$

Equations (4.2) and (4.3) are in fact local balance equations that sum up to give the global balance equation Eq. (4.1).

One can imagine how tedious this can be if there are more jobs (so there are more states) or more queues (so there are more dimensions). In contrast, an open separable network can be easily solved using Jackson's Theorem (Sec. 3.3), even though it has infinitely many states.

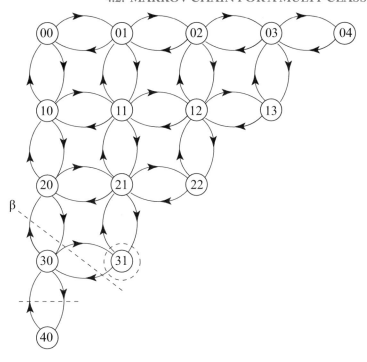

Figure 4.3: One can derive equations for solving the Markov chain by balancing flows across boundaries.

## 4.2    MARKOV CHAIN FOR A MULTI-CLASS NETWORK

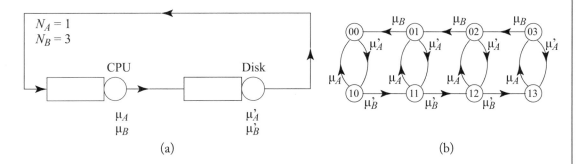

Figure 4.4: Class $A$ has preemptive priority over class $B$ in (a); (b) is the corresponding Markov chain for state $n_A n_B$ at the CPU.

Now consider a closed network with two queues—for CPU and disk—and two job classes. Class $A$ has **preemptive** priority over class $B$ at both queues, i.e., whenever a class $A$ job enters a queue, any class $B$ job that is in service will be suspended, to let the class $A$ job begin service immediately.

If the servers are memoryless, then such suspension does not affect the service times for class $B$ jobs, but their waiting times are obviously affected by class $A$. On the other hand, class $B$ is "invisible" to class $A$ jobs.

Suppose there are 1 class $A$ job and 3 class $B$ jobs, and service rates are $\mu_A$ and $\mu_B$ at the CPU queue and $\mu'_A$ and $\mu'_B$ at the disk queue—see Fig. 4.4a. If $n_A n_B$ denotes $n_A$ class $A$ jobs and $n_B$ class $B$ jobs at the CPU, we again get a 2D Markov chain, as shown in Fig. 4.4b.

## 4.3    STATE AGGREGATION

To overcome the intractability of a large, multidimensional Markov chain, one can resort to approximation by **state aggregation**. To illustrate, we can group all states in Fig. 4.2 with $n_B + n_C = k$ into one state $\{k\}$, like in Fig. 4.5a, and thus collapse the 2D chain into the birth-death process in Fig. 4.5b.

However, what should the aggregated transition rates be? We can interpret Fig. 4.5b as the Markov chain for a two-queue network. In effect, we are aggregating queues $B$ and $C$ in Fig. 4.1 into one queue $D$. One could possibly approximate the "service rate" for this aggregate queue as $\mu_D = p\mu_B + (1-p)\mu_C$. We then get the network in Fig. 4.6a; the corresponding Markov chain is Fig. 4.5b, with the transition rates in Fig. 4.6b. The solution to this Markov chain gives $\text{Prob}(n_B + n_C = k)$ (or, equivalently, $\text{Prob}(n_A = 4 - k)$) for $k = 0, 1, 2, 3, 4$ without providing approximations for $\text{Prob}(n_B = i \text{ and } n_C = j)$.

There is a flaw in this approximation: while $\mu_D = p\mu_B + (1-p)\mu_C$ properly aggregates the service time, it does not correctly aggregate the queueing performance. To see this, consider two possibilities: (i) $p = 0.2$, $\mu_B = 3$, and $\mu_C = 8$ and (ii) $p = 0.5$, $\mu_B = 7 = \mu_C$. In both cases, $\mu_D = 7$, and their solution for Fig. 4.6b is the same. Now, let $X$ be the throughput. For single-server queues, utilization is bounded by 1, so we get $\frac{0.2X}{3} < 1$ and $\frac{0.8X}{8} < 1$ for (i) and $\frac{0.5X}{7} < 1$ for (ii). Therefore, $X < 10$ for (i) and $X < 14$ for (ii), so the performance for the two cases are in fact different.

We will return to this example in Sec. 6.2, where we will see a more rigorous way of aggregating the two queues.

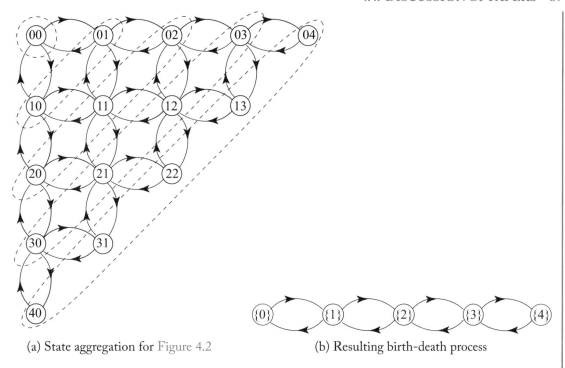

(a) State aggregation for Figure 4.2

(b) Resulting birth-death process

Figure 4.5: Approximate solution of a Markov chain by state aggregation, where $\{k\} = \{n_B n_C \mid n_B + n_C = k\}$.

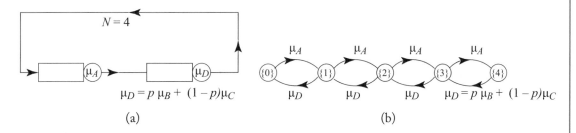

(a)

(b)

Figure 4.6: A plausible alternative: Queues $B$ and $C$ in Fig. 4.1 aggregated into queue $D$ in (a); (b) shows the corresponding Markov chain.

## 4.4   DISCUSSION OF PAPERS

Markov chains are powerful—they can even be used to analyze the correctness of mutual exclusion algorithms [12]. In the following, we examine the Markov chain for *GPRS* [42], then introduce another three papers with Markov chains (*SensorNet* [49], *DependabilitySecurity* [61], and *ServerEnergy* [19]).

### GPRS [42]

We see the use of aggregation to estimate $r^i(n)$, the probability that an inactive mobile in cell $i$ that wants to start a new transfer cannot do it because the system limit $M_{\max}$ is reached, assuming that there are $n$ mobiles currently in active transfer in the cell.

If $\langle a, b \rangle$ denotes a state in *Fig. 4*, then each diagonal with $a + b = k$ is collapsed into an aggregated state with probability

$$p_{uc}^{S^i}(k) = \mathrm{Prob}(\langle 0, k \rangle) + \mathrm{Prob}(\langle 1, k - 1 \rangle) + \cdots + \mathrm{Prob}(\langle k - 1, 1 \rangle) + \mathrm{Prob}(\langle k, 0 \rangle).$$

If one uses the approximation $\mathrm{Prob}(\langle a, b \rangle) \approx p_{uc}^j(a) p_{uc}^j(b)$, then

$$p_{uc}^{S^i}(k) \approx p_{uc}^j(0) p_{uc}^j(k) + p_{uc}^j(1) p_{uc}^j(k - 1) + \cdots + p_{uc}^j(k - 1) p_{uc}^j(1) + p_{uc}^j(k) p_{uc}^j(0),$$

which is the convolution in *Eq. (13)*.

The probability $r^i(n)$ is thus approximated in *Eq. (14)* as the ratio of the aggregated probability for the last diagonal to the probability for the triangle in *Fig. 4*.

### SensorNet [49]

The system of sensors, MULEs and access points is illustrated in *Fig. 1*. The basic issues are how buffer sizes on sensors and MULEs should scale and how delivery rates from sensors to access points (APs) would scale with the system size. Related issues are delivery latency, network longevity, MULE failure, data loss, and retransmissions.

The major input parameters listed in *Sec.IV* are grid size $N$ and number of APs, MULEs, and sensors ($N_{AP}$, $N_{mules}$, $N_{sensors}$). The metric of interest is delivery rate $S$; buffer sizes, for example, are only studied through simulation (*Table IV*).

The authors adopt a closed model ($N_{mules}$ is fixed) that discretizes both time and space: APs and MULEs are located at grid points; sensors generate data and MULEs move at clock ticks. Data is transferred from sensor to MULE only when they are at the same grid point.

The random variable $Y_k$ indicates if one or more MULEs intersect the sensor at time $k$, with probability $P(Y_k = 1) = 1 - (1 - 1/N)^{N_{mules}}$. This expression includes the probability that two or more MULEs are at the same position at the same time, which is an artifact of the space/time discretization.

One could avoid such an artifact by using continuous time and space, and introducing a transmission radius around each sensor (or each MULE): data is transferred to the first MULE to come within this distance of a sensor. There is a possibility that a MULE may come within reach of two sensors, but the model can assume that transmission radii are tuned to avoid such an overlap (consistent with the omission of wireless interference from this model). However, one may still have to calculate the probability that one or more MULEs are simultaneously within reach of a sensor. (When computing the hitting time distribution in *Sec. VII*, the model does use continuous time.)

To study how buffer sizes should scale, the model makes them infinite, and derive expressions for their expected occupancy (*Eq. (5)*).

Like *GPRS* [42], the model is a discrete-time Markov chain with transition probabilities between states (instead of transition rates in the case of continuous-time Markov chains).

*DependabilitySecurity* [61]

There are two reasons for including this paper in our discussion: (i) it illustrates how one can construct quantitative models for dependability and security (instead of performance); and (ii) it gives many examples of Markov chains.

It is sometimes hard to separate performance from dependability and security. A disk failure may reroute requests to a remote copy and thus increase latencies, while a typical denial-of-service attack tries to overload a system. It is therefore natural for the authors to propose a unifying framework for classifying performance, dependability, and security.

Since the point of this paper is to illustrate the framework in various contexts, it does not look at one particular system, and the model parameters and performance metrics vary with the system under consideration.

The framework includes combinatorial models, like *Fig. 5* for reliability of virtual machines, *Fig. 10* for availability of BGP sessions, and *Fig. 17* for security of smart cards. Often, a system attribute can be modeled in two different ways; for example, network security can be modeled with a reliability graph (*Fig. 6*) or fault tree (*Fig. 7*).

Conversely, one model can serve multiple purposes. The birth-death process in *Fig. 11* is the performance model for the $M/M/K/K$ queue in Fig. 3.5, but is similar to the availability model in *Fig. 8* (except "arrivals" are failures, and "service times" are repair times).

These two examples, *Fig. 11* for performance and *Fig. 8* for availability, are combined into the two-dimensional Markov chain in *Fig. 14* that simultaneously models performance degradation and hardware failures. The ease of such a combination illustrates the flexibility of Markov chains.

*ServerEnergy* [19]

ACES aims to balance the performance, cost, and reliability of servers in a data center. The model discretizes time into steps of equal length, then converts client request rates into $m_j$, the number of servers needed to satisfy the requests at step $j$. The objective is thus a transient analysis of the system, rather than the steady state solution of the model.

ACES turns servers on and off to meet the *demand* $m_j$: If $On_j$ is the number of servers that are turned on, and $On_j < m_j$, then there is a performance loss and possibly a violation of Service Level Objective (SLO); if $On_j > m_j$, then there is excess energy consumption. ACES first minimizes $E = \sum_j |m_j - On_j|$ to get a minimum value $E_C$ for each $C$, where $C$ factors in reliability by modeling the cost of transition between power states; it then determines the $C$ value that minimizes $E_C$.

Booting up a server can take a while, and waiting for workload to drain (e.g., users logging off so a server can be powered down) can take even longer. Transitions between power states can therefore span multiple time steps. ACES must therefore anticipate demand $\hat{m}_0$, and it does so by taking a linear combination of past demand (*Eq. (1)*), $\hat{m}_0 = a_p m_{-p} + \cdots + a_1 m_{-1} = \boldsymbol{a} \boldsymbol{m}_{-1}$. This $\boldsymbol{a}$ is determined by linear regression using $r$ samples $\mathbf{m}_{-r}, \ldots, \mathbf{m}_{-1}$ that are weighted by $w_{-i} = \alpha^{i-1}$ ($\alpha = 0.95$). In contrast, the prediction in *ProactiveReplication* [13] is based on a single previous period $\Delta T$.

The (power) state diagram for a server (*Fig. 4*) is a variation on a Markov chain: although the transition times can be converted into transition rates, the choice of transitions and state residence times are controlled by ACES. The use of equal time steps allows the model to unroll the transition among states, where a "state" in *Fig. 6* is now the number of servers in a particular power state. The usual local/global balance equations for transition rates in a Markov chain are now discretized, to give server conservation equations for each time step (*Eq. (3)*).

CHAPTER 5

# Closed Systems

As we can see from the previous chapter, the solution of closed queueing networks can be very tedious. This is so even for separable networks (which are easy to solve if they are open, as we saw in Chapter 3).

The solution is easier if we give up state probabilities, and only solve for average values. This chapter starts with the PASTA property that gives us the Arrival Theorem, which is the basis for the MVA solution algorithm.

## 5.1    POISSON ARRIVALS SEE TIME AVERAGES (PASTA)

Recall from Chapter 3 that, for a time sequence $T_1, T_2, \ldots$, a random arrival at time $t$ within an interval $(T_{n-1}, T_n)$ may see $E(T_n - T_{n-1})$ different from $E(T_i - T_{i-1})$ seen by an outside observer.

Now consider a system with Poisson arrivals, and each arrival observes some variable $v$ in the system state upon arrival. A well-known property says that the average $v$ observed by the arrivals is the same as that seen over time by an outside observer. This property is called **PASTA**, for Poisson Arrivals See Time Averages.

**Example:** $M/M/1$

An $M/M/1$ queue has Poisson arrivals, and the average queue length over time is $\bar{n} = \frac{\rho}{1-\rho}$ from Eq. (2.5); by the PASTA property, this is also the average queue length seen by an arrival. Since service times are memoryless, the residual service time of the job in service is $\frac{1}{\mu}$, so the average waiting time for an arrival should be

$$W = \bar{n}\left(\frac{1}{\mu}\right) = \frac{\rho}{1-\rho}\frac{1}{\mu},$$

as in Eq. (2.8).

**Example:** $M/D/1$

For $M/D/1$, although the average queue length seen by the Poisson arrivals is again the time average $\bar{n}$, we no longer have $W = \bar{n}\left(\frac{1}{\mu}\right)$. This is because the deterministic service time is not memoryless, so the residual life of the job in service when a job arrives is no longer $\frac{1}{\mu}$. To confirm, Little's Law and Eq. (2.9) give

$$\bar{n} = \lambda\left(W + \frac{1}{\mu}\right), \quad \text{so } W = \frac{\bar{n}}{\lambda} - \frac{1}{\mu} = \left(1 - \frac{\rho}{2}\right)\frac{1}{1-\rho}\frac{1}{\mu} - \frac{1}{\mu} = \frac{\rho}{2(1-\rho)}\frac{1}{\mu}, \tag{5.1}$$

which is different from $\bar{n}\left(\frac{1}{\mu}\right) = \frac{(2-\rho)\rho}{2(1-\rho)}\frac{1}{\mu}$. To derive $W$ using PASTA, note that an arrival sees $\rho$ jobs in service. Since the residual life is $\frac{1}{2\mu}$ (Sec. 3.1), we get

$$W = (\bar{n} - \rho)\frac{1}{\mu} + \rho\frac{1}{2\mu} = \frac{\rho}{2(1-\rho)}\frac{1}{\mu},$$

in agreement with Eq. (5.1).                                                                $\square$

## 5.2   ARRIVAL THEOREM

Separability has been extended to closed networks. For example, the network can have single servers with exponentially distributed service times using FCFS or processor sharing, as well as delay centers (infinite server queues). Such networks are again separable, with a product-form joint distribution

$$\text{Prob}(n_1 = k_1, \ldots, n_M = k_M) = \frac{1}{G(M, N)}f_1(k_1)\cdots f_M(k_M), \tag{5.2}$$

where $M$ is the number of queues, $N$ is the number of jobs, $G(M, N)$ is a normalization constant (that makes the right-hand side of Eq. (5.2) less than 1), and each $f_i(k_i)$ depends on properties of queue $i$. Equation (5.2) can be further generalized to multiclass networks [23].

The computational complexity in solving Eq. (5.2) is dominated by the need to evaluate $G(M, N)$. One can get around this difficulty by evaluating just the average performance (instead of the detailed probabilities). The idea for doing so capitalizes on PASTA.

For a closed network, the arrivals at a queue are correlated with the network state. For example, if there is only one job, then that job will only see empty queues, whereas the time-averaged queue sizes (seen by an outside observer) are nonzero. Nonetheless, the **Arrival Theorem** [33, 48] says that, for a separable closed network with $N$ jobs, an arrival at a queue sees a network state that is (distribution-wise) the same as that seen by an outside observer of the same network with $N - 1$ jobs.

# 5.3    MEAN VALUE ANALYSIS (MVA)

We now describe how the Arrival Theorem can be used for solving closed networks. We need the following notation.

$$
\begin{aligned}
M &= \text{number of queues} \\
\delta_i &= \begin{cases} 0 & \text{if queue } i \text{ is a delay center} \quad (\text{Sec. } 2.2) \\ 1 & \text{if queue } i \text{ has 1 server} \end{cases} \\
N &= \text{number of jobs in the network} \\
V_i &= \text{average number of visits a job makes to queue } i \\
S_i &= \text{average service time at queue } i \text{ for a job} \\
D_i &= V_i S_i \text{ (called \textbf{service demand} at queue } i) \\
Q_i(N) &= \text{average number of jobs at queue } i \\
A_i(N) &= \text{average number of jobs seen by an arrival at queue } i \\
R_i(N) &= \text{average time per visit a job spends at queue } i \\
X(N) &= \text{throughput}
\end{aligned}
$$

It follows from the Arrival Theorem that

$$
A_i(N) = Q_i(N - 1) \quad \text{for every queue } i, \tag{5.3}
$$

i.e., the average queue length seen by an arriving job $\mathcal{X}$ is the same as the time-averaged queue length seen by an outside observer—with $\mathcal{X}$ removed. This is an intuitive result that is analogous to PASTA for open systems.

Now, the time a job spends at queue $i$ per visit is the sum of its waiting and service times. Therefore, if all queues are either delay centers or one-server queues, and their service times are memoryless, then

$$
R_i(N) = \delta_i A_i(N) S_i + S_i = S_i(1 + \delta_i Q_i(N - 1)). \tag{5.4}
$$

In steady state, the throughput at queue $i$ is $V_i X(N)$ (this is sometimes called **Forced Flow Law** [10], from operational analysis), so

$$
\begin{aligned}
Q_i(N) &= V_i X(N) R_i(N) \quad \text{by Little's Law} \\
&= D_i(1 + \delta_i Q_i(N - 1)) X(N). \tag{5.5}
\end{aligned}
$$

The above derivation only works with the average values, so it is called **Mean Value Analysis (MVA)**. It has been extended to include other types of queues and multiple job classes.

The average total time a job spends in the system is $\sum_{i=1}^{M} V_i R_i(N)$, so by Little's Law

$$X(N) = \frac{N}{\sum_{i=1}^{M} V_i R_i(N)} = \frac{N}{\sum_{i=1}^{M} D_i(1 + \delta_i Q_i(N-1))} \qquad \text{from Eq. (5.4).}$$

Using this and the initialization $Q_i(0) = 0$, Eq. (5.5) provides a recursive MVA algorithm for solving the network. Notice that this solution requires only service demand $D_i$, and does not need to know $V_i$ and $S_i$.

One can skip the recursion with **Schweitzer's Approximation** [47]

$$Q_i(N-1) \approx \frac{N-1}{N} Q_i(N).$$

Substituting this approximation into Eq. (5.5), we get a **fixed-point approximation** in $Q_i(N)$, which can be solved iteratively, as follows:

$$
\begin{aligned}
Q_i(N) \quad &\leftarrow \quad \frac{N}{M} \qquad \text{//initialization} \\
\text{repeat}\{ \\
X(N) \quad &\leftarrow \quad \frac{N}{\sum_i D_i(1 + \delta_i \frac{N-1}{N} Q_i(N))} \\
Q_i(N) \quad &\leftarrow \quad D_i(1 + \delta_i \frac{N-1}{N} Q_i(N))X(N) \\
\}\text{until convergence.}
\end{aligned}
$$

In practice, this iterative approximation is usually faster than the recursive exact solution, and may suffice in terms accuracy.

## 5.4   DISCUSSION OF PAPERS

Below, we discuss four examples (*InternetServices* [63], *NetworkProcessor* [34], *PipelineParallelism* [41], and *DatabaseScalability* [14]) of using a closed queueing network to model system performance.

*InternetServices* [63]

The main issue for this paper is in modeling the multiple tiers that are traversed by an application (*Fig. 1*), taking into account sessions, caching, and concurrency limits, so the model can be used for capacity provisioning, bottleneck identification, session policing, etc.

The model chosen is an interactive (closed) queueing network (*Fig. 3* and *Fig. 5*). The main input parameters are $M$ (number of tiers), $N$ (number of sessions), $\bar{Z}$ (average think time) and service demand $\bar{D}_m$; the main performance measures are throughput $\tau$ and average response time $\bar{R}$ (see *Table 1*). Note that, making $\bar{Z}$ an input parameter implicitly assumes that user behavior is independent of response times—this may not be so.

Sessions are modeled by delay centers, caching is modeled by the loopbacks from tier $Q_i$ to tier $Q_{i-1}$, and concurrency limits are modeled by $M/M/1/K_i$ queues in *Step 2* of the solution algorithm in *Sec. 4*; *Step 1* is a straightforward MVA algorithm.

Note that the approximation $V_i \approx \frac{\lambda_i}{\lambda_{\text{request}}}$ in *Sec. 3.3* would not work well in *Sec. 4*, where requests can be dropped by the concurrency limit.

### *NetworkProcessor* [34]

This paper gives an example of how an analytical model can be used to explore a design space, namely the implementation of some software. Specifically, they consider the implementation of a network application on a given network processor.

Without help from an operating system, the implementation must itself manage the resources (e.g., cycles, threads, memory) at the packet processors. An implementation decides its data flow through the components, and thus its performance. An analytical model facilitates a comparison of implementation possibilities, thus reducing implementation effort. Unfortunately, *Sec. IV* does not illustrate the use of the model to compare different implementation possibilities.

*Fig. 1* shows the architecture of the network processor but, since the issue is resource management at the packet processors, the analytical model in *Fig. 2* focuses on the packet processors and memory. By using a queueing network, the model captures the data flow of the software implementation through the hardware.

Usually, service demand $D_i$ for a queueing network model are obtained by measurements (e.g., with microbenchmarks). The technique of setting $D_i$ through instruction counts is an interesting alternative, but one imagines that this can get hairy if the code has multiple conditional branches.

The authors adopt an open model by fixing the packet arrival rate. However, with finite buffer sizes, some packets are lost if arrival rate is high. A network application (e.g., TCP) may react to the loss by changing its transmission rate (e.g., change the congestion window and retransmit lost packets). In other words, some applications may force one to model a feedback loop that alters the packet arrival rate.

The decomposition of an $m$-server queue into single-server and delay center (*Fig. 3*) is similar to the Schweitzer Approximation. This decomposition makes it possible to use the MVA equations.

The queueing network in *Fig. 2* is partly solved by using MVA. *Eq. (3-5)* has a variable *J* for the number of requests when throughput reaches maximum in the subnetwork; it is not clear what this *J* refers to, since there is no such maximum for a closed separable network.

*PipelineParallelism* [41]

This paper illustrates how an analytical model can be used to find a good way of structuring a parallel computation. The execution is a pipeline that is divided into stages, with multiple threads per stage. The central issue is how to balance the computation over this pipeline structure, so no bubble forms at some stage and thus stall the other stages. The authors consider two ways of preventing bubbles: (i) collapsing stages so each stage has more threads and (ii) having idle threads steal work from overloaded busy threads.

The pipeline structure is determined by two key parameters—the number of stages and the total number of threads. Performance is measured by workload execution time, speedup (ratio of execution time for *p* cores to that for 1 core), and efficiency (speedup divided by *p*).

The analysis shows that the performance model for a parallel execution can depend on how the code is implemented. The Pthreads PARSEC implementation of `ferret` and `dedup` are both modeled with one queue per stage. However, *Sec. 3.1* uses a closed queueing network to model the stalling caused by finite buffers in the `ferret` implementation (see the last stage in *Fig. 2*), but *Sec. 3.2* uses an open network for `dedup` because, in that case, the buffers are effectively infinite.

The model can also depend on how the execution is scheduled. Whereas the Pthreads implementation is modeled with one queue per stage (*Sec. 3.2*), the TBB implementation that is executed with work stealing is modeled with one queue per thread.

For an open model, the throughput is equal to the arrival rate, which is an input parameter, instead of calculated ($\lambda_e^{st5}$ and $\lambda_e^{st3}$ in *Sec. 3.2*). Since the authors assume arrivals are Poisson and service times are exponential, the execution time in *Fig. 6*) can be determined by using an extension of Jackson's Theorem (Sec. 3.3) to separate the network in *Fig. 7* into isolated queues and calculating the system time for each. Note that *Fig. 7* shows each queue feeding into the next; this is misleading since some items bypass some stages—see *Fig. 3*—which is why $T_{arr_i}$ varies with $i$.

*DatabaseScalability* [14]

This paper illustrates how a model can use the performance of a standalone database to predict the performance of a replicated database and, in addition, compare two alternative designs (multi-master and single-master). The components of these systems are shown in *Fig. 1* and *Fig. 2*; in addition, there is a mix of read-only and update transactions. With snapshot isolation, the read-only transactions do not conflict with others, so the only conflicts are between transactions that write.

The values for some of the input parameters are extracted from the log file (e.g., $Pr$), or by measurement (e.g., $A'_N$) on a standalone system.

Instead of directly applying MVA to the $N$-replica system (*Fig. 1*), the authors apply it to each replica individually. In effect, this is hierarchical decomposition, but the different subnetworks (i.e., replicas) affect each other through the workload generated by the abort of update transactions. This is why the demand $D_{MM}(N)$ includes $N$ and abort rate $A_N$.

When clients are added in the MVA iteration, the read-only and update transaction rates would increase, so $R$ and $W$ are not fixed, like in an open model.

Aborted transactions are resubmitted, so the update transaction rate is scaled up to $W/(1 - A_N)$ by using the geometric distribution (Lemma 0.7), assuming a resubmitted transaction has the same abort probability. One could model this effect by similarly scaling the number of visits each transaction makes to each queue; the $Pw/(1 - A_N)$ factor in demand $D_{MM}(N)$ then follows.

The model's use of standalone profiling to predict trasaction conflict for a multi-master system rests on the formula $(1 - A_N) = (1 - A_1)^{\frac{NCW(N)}{L(1)}}$. As conflict window $CW(N)$ includes transacion execution time and thus queueing delays, the model progressively changes $CW(N)$ as the number of clients increase in the MVA iteration. This also means $A_N$ increases as clients are added, so the mix of transactions is not a constant $Pr : Pw$, but shifts toward updates. Moreover, MVA assumes service demand does not change as more jobs are added, whereas $D_{MM}(N)$ changes when $A_N$ increases with the number of clients.

Note the application of Little's Law to estimate the number of concurrent update operations ($CW(N) \cdot N \cdot W \cdot U$). Strictly speaking, however, having updates means the database will be constantly changing, so there is no steady state and we cannot apply MVA.

# CHAPTER 6

# Bottlenecks and Flow Equivalence

We see that, for a separable network, we can compute the joint distribution with Jackson's Theorem if it is open, and compute the mean performance measures with MVA if it is closed. Although the separability requirement has been progressively relaxed, it remains restrictive. For example, although one can view the Internet as a massive queueing network, there are few queueing-theoretic models of the Internet; the finite buffers in routers, the bounded packet lengths, the feedback effect of packet acknowledgements, etc. all get in the way of applying results from queueing theory.

There are no general techniques for solving nonseparable queueing networks. Even so, we can often get bounds by analyzing bottleneck queues, or get an approximate solution by decomposing the network, as illustrated in this chapter.

## 6.1 BOTTLENECK ANALYSIS

Consider a (closed or open) queueing network, with throughput $X$. Suppose queue $i$ has $K_i$ servers, with average service time $S_i$ per visit. Let $V_i$ be the average number of visits to queue $i$ per job, and service demand $D_i = V_i S_i$. By Little's Law, the utilization per server is

$$\frac{V_i X S_i}{K_i} = \frac{X D_i}{K_i} \leq 1, \quad \text{so } X \leq \frac{K_i}{D_i}.$$

This is true for every $i$, so

$$X \leq \min_i \frac{K_i}{D_i}.$$

A queue that gives the minimum $K_i/D_i$ or (equivalently) the maximum service demand per server $D_i/K_i$ is called a **bottleneck**.

A bottleneck also maximizes $X D_i/K_i$, so it has highest utilization per server. There is always a maximum $D_i/K_i$, so a bottleneck always exists. Relieving a bottleneck by speeding up its servers (i.e., reducing $D_i$) or adding servers (i.e., increasing $K_i$) only shifts the bottleneck to another queue.

Note that a bottleneck can be identified from $K_i$ and $D_i$, without knowing $V_i$ and $S_i$, nor solving for $X$. Identifying a bottleneck in a network is important; as its name suggests, a bottleneck defines a bound and, to a large extent, it determines the performance of a system. This is one reason many simulation and analytical studies of Internet traffic adopts a dumbbell model, like in Fig. 6.1. A bottleneck may not be caused by hardware; for example, TCP's congestion window limits a connection's throughput, so it can act like a bottleneck.

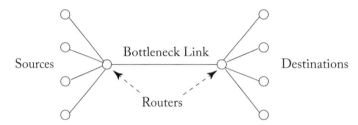

Figure 6.1: The dumbbell model frequently used in Internet traffic analysis.

For an open queueing network, a bottleneck defines a capacity bound, like the vertical asymptote in Fig. 2.5 for an $M/G/1$ queue—performance deteriorates rapidly as workload approaches that bound.

For a closed network, Little's Law gives $N = XT$, where $T$ is system time. Now, when the population size $N$ is small, there is little queueing, so $T \approx \sum_i D_i$; therefore, $X \approx \frac{N}{\sum_i D_i}$, which is approximately linear in $N$.

For large $N$, $X$ approaches the bound $\min_i \frac{K_i}{D_i}$ imposed by the bottleneck, so

$$T \approx \frac{N}{\min_i \frac{K_i}{D_i}}, \quad \text{which is approximately linear in } N.$$

The shapes of $X$ and $T$ as functions of $N$ are thus as shown in Fig. 6.2; in particular, a bottleneck defines a throughput barrier on $X$. Note that the asymptotically linear increase in $T$ for a closed system is significantly different from the super-linear increase for an open system like the $M/G/1$ in Fig. 2.5. The difference lies in the feedback for a closed system: If a system is open, new jobs arrive regardless of congestion, so completion time balloons; if the system is closed, however, then a slow down in job completion time also slows down the generation of new jobs, thus moderating the increase. For the interactive (closed) system in Fig. 1.4, we get corresponding bounds on response time $R$ from $R = T - Z$.

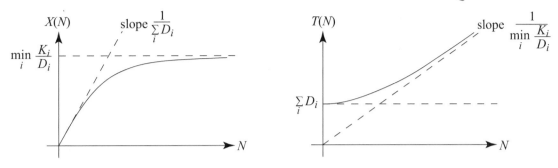

Figure 6.2: For a closed network, the shapes of throughput $X$ and system time $T$ for large $N$ are determined by bottlenecks [39].

Bottleneck analysis is powerful, in that it requires no assumptions about probability distributions and queueing network topology. On the other hand, it is weak in that it provides no information on the gap between the bounds and the actual performance. This gap is largest around the **knee** where the two bounds intersect.

Note from Fig. 6.2 that the knee is determined by maximum throughput $X_\infty = \min_i \frac{K_i}{D_i}$ and minimum delay $T_0 = \sum_i D_i$, so it is intuitively where performance is "optimal." This observation is the basis for a recent variant of TCP [5], where $X_\infty$ is the network's bottleneck bandwidth, $T_0$ is the minimum round-trip time, and the knee $X_\infty T_0$ is the *bandwidth-delay product*. Using the idea underlying Schweitzer's Approximation (Sec. 5.3), one can estimate $X_\infty$ and $T_0$ asymptotically [54].

Recall the calculation in Sec. 4.3 that shows the queue aggregation in Fig. 4.6a is flawed; that calculation is in fact an application of bottleneck analysis. We next see how the aggregation can be done in a rigorous way.

## 6.2    FLOW EQUIVALENCE

Recall that, for an $M/M/3$ queue, the service rate depends on the number of jobs $k$ in the queue. This is an example of a **load-dependent** queue, represented as

with output rates $X(1) = \mu$, $X(2) = 2\mu$, and $X(k) = 3\mu$ for $k \geq 3$.

Load-dependent queues are common in computer systems. Consider these examples: The service time at a CPU may increase with the number of processes because of context switching; the retrieval time from a cache shared by $N$ jobs will increase with $N$ because of cache contention and misses; seek time and rotational latency can decrease as a disk queue grows; and an increase in the number of clients in a wireless cell can cause packet collisions and retransmissions, and thus slow down the base station. In each case, the output rate $X(N)$ is nonlinear in $N$.

We saw in Chapter 4 the computational tedium in solving a multidimensional Markov chain, and the idea of deriving a faster approximation through state aggregation. One way of doing this aggregation is to isolate and solve a subsystem (this is called **hierarchical decomposition**), plug it back as a load-dependent queue (i.e., **aggregation** of the subsystem), then solve the aggregated system (via **flow equivalence**).

For example, the three-queue closed network in Fig. 4.1 can be decomposed by aggregating queues $B$ and $C$ into one load-dependent queue that represents a subnetwork, like in Fig. 6.3. The subsystem $D$ can be solved by MVA for all $K$, or by a Markov chain for each $K$, to give $X_D(1)$, $X_D(2)$, $X_D(3)$, and $X_D(4)$. The aggregated system can then be solved by another Markov chain for state $K$ in the subsystem $D$.

The idea is: The flow through the load-dependent queue $D$ is equivalent to flow around the subnetwork. This flow equivalence is in fact exact for separable networks. Thus, the right way of calculating the transition rates for the aggregated queue is Fig. 6.4, not Fig. 4.6b. However, if all we want is throughput (and not state probabilities) then we might as well solve the three-queue network itself with MVA (Sec. 5.3).

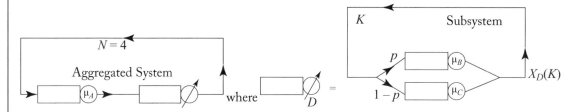

Figure 6.3: Aggregating a subnetwork in Fig. 4.1 into a load-dependent queue.

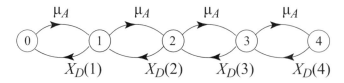

Figure 6.4: Correct transition rates for the aggregated queue in Fig. 4.6a.

The real use of flow equivalence is for nonseparable networks (as when the load-dependent queue models, say, a cache hierarchy or a wireless cell), for which it is an approximate technique. This approximation is good if the subnetwork reaches steady state faster than the complementary network [9]. Intuitively, every arrival to the subnetwork sees it in equilibrium with a steady $K$ value.

## 6.3    EQUIVALENCE BETWEEN OPEN AND CLOSED

Sometimes, a closed system is easier to analyze, e.g., there are a fixed number of threads in a parallel computation, or a fixed number of transmitters in a wireless cell. In that case, one can approximate an open system by flow equivalence with a closed system, as illustrated in Fig. 6.5.

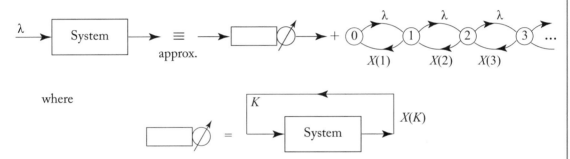

Figure 6.5: An open system can be approximated with a closed network as a load-dependent queue.

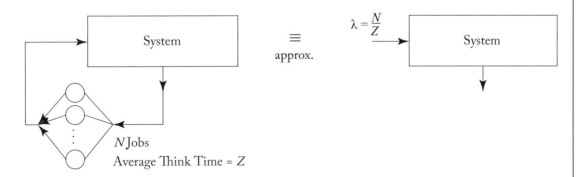

Figure 6.6: A closed interactive system can be approximated as an open system if $N$ is large.

Conversely, a closed interactive system can be approximated by an open system, as illustrated in Fig. 6.6, e.g., requests arriving at a website are usually modeled with an open system. If there are $K$ jobs in the system, then there are $N - K$ users (at the delay server) that are each generating a job at rate $1/Z$. For $N >> K$, the arrival rate to the open system can be approximated as $N/Z$. For separable networks, if the think time $Z$ is scaled so that $\lim_{N\to\infty} \frac{N}{Z} = \lambda$ for some constant $\lambda$, then the two models are in fact equivalent in the limit [32].

## 6.4   DISCUSSION OF PAPERS

We now discuss the bottlenecks and flow equivalence in the previous papers, as well as four others: *OpenClosed* [46], *DatabaseSerializability* [3], *Roofline* [66], and *PerformanceAssurance* [45].

*InternetServices* [63]

Since the paper uses a closed interactive model, the response time is $R = T - Z$, where $T$ should have the shape in Fig. 6.2. However, the response time curve for the baseline model in *Fig. 4(a)* looks different, but a closer look shows a small nonlinear curve near the origin. In other words, *Fig. 4(a)* shows mostly the linear increase imposed by the bottleneck in the basic model. (See also the response and residence times in subsequent plots.) If one is interested in this bottleneck constraint, then there is no need for the elaborate model in *Fig. 3*.

Similarly, *Fig. 4(b)* shows good agreement between the enhanced model and the measured response time in the high load region where the concurrency limit is in effect. Here, the model is helped by the assumption that connections are turned away when the concurrency limit is reached; if these rejected requests were buffered and served later, modeling the response time accurately may be harder.

*StorageAvailability* [18]

Bottlenecks in a system may be caused not by hardware but by software. *Sec. 7* gives an example of such a bottleneck: a lock held by the HE causes writes to be serialized.

*NetworkProcessor* [34]

Although the overall model is open (with packet arrival rate as an input parameter), this is later solved through flow equivalence with a closed queueing subnetwork. Except for the variable $J$, *Eqs. (3-10)–(3-15)* are equations obtained by solving the Markov chain for Fig. 6.5.

The network can be solved directly (without using flow equivalence) by using Jackson's Theorem for open networks. This works with multiserver queues, without need for the queue decomposition in *Fig. 3*; however, it is more compute-intensive, since the solution works out the entire joint distribution of queue sizes, instead of just the average queue sizes.

The paper also uses bottleneck analysis (*Eq. (3-4)*) to make the MVA algorithm converge. This possibly refers to setting a threshold for stopping the MVA iteration (when calculated throughput is sufficiently close to the throughput bound set by the bottleneck).

*RouterBuffer* [1]

This paper gives an example of the idea in Fig. 6.5: the router analysis is in terms of $n$, the number of flows. In effect, a closed model is used to analyze an open system.

*DependabilitySecurity* [61]

In flow equivalence (Sec. 6.2), we extract a subnetwork of queues, solve it, plug it back as one load-dependent queue, then solve the bigger network with a birth-death process with rates $\mu_A$ and $X_D(n)$. This idea of hierarchical decomposition extends beyond queueing networks. In the virtualization reliability model (*Fig. 5*) and router availability model (*Fig. 9*), each block may represent, say, a separate Markov chain. Moreover, in the case of *Fig. 9*, the metric is not performance, but **dependability**.

Sometimes, a submodel may have a parameter whose value is known only after the bigger model is solved. For example, the Markov chain in the submodel has an arrival rate that is determined by the throughput of the bigger model. One can deal with such circularity with an iterative solution: Initialize the throughput somehow; the submodel uses the corresponding arrival rate for a solution; this submodel solution is used to solve the bigger model and improve the throughput estimate; then the solutions iterate until convergence. This is an example of the fixed-point iterative model (see *Table 1*).

*PipelineParallelism* [41]

*Figs. 2* and *3* illustrate how the bottleneck in a system can shift—in this case, from an intermediate stage to an IO stage as the number of threads increases. Note the bottleneck analysis in *Eqs. (7), (9)*, and *(10)*.

*Sec.3.1* calculates the mean inter-arrival time as $T_{arr} = \max(T_{ser_i})$. This assumes that the number of items $K$ is so large that the bottleneck queue ($\max(T_{ser_i})$) is never idle.

This $T_{arr}$ is then used as the service time for each server in a delay center that feeds into the $c$-server queue in *Fig. 5(b)*. In effect, the authors are using hierarchical decomposition, with all queues except one in *Fig. 5(a)* aggregated as a load-dependent delay center in *Fig. 5(b)*.

*DatabaseScalability* [14]

*Table 1* suggests read-only transaction rate $R$ and update transaction rate $W$ are inputs, like in an open model. However, the queueing networks in *Fig. 1* and *Fig. 2* are closed, so $R$ and $W$ should be determined by the MVA solution. Alternatively, one could view $R$ and $W$ as the limiting rates when the number of transactions is large and the closed networks behave like open systems (Fig. 6.6).

Except for *Fig. 6*, the throughput and response time plots from the experiments look just like those determined by bottlenecks in Fig. 6.2. (By increasing the number of replicas in *Fig. 6*, the experiments also increase the number of clients.) In other words, one could get these plots from the service demands without the need for MVA and conflict analysis.

*OpenClosed* [46]

For both analytic and simulation models, whether it is open or closed can have significant impact on its accuracy and conclusions.

*Principle (i)* and *Principle (iii)* intuitively follows from the fact that, unlike in an open system, every queue in a closed system has a maximum size. The convergence of a closed system to an open system stated in *Principle (ii)* is similar to the result mentioned in Sec. 6.3.

One can understand *Principle (iv)* and *Principle (v)* as follows: From Little's Law, $\overline{n} = \lambda T$ implies that, in an open system where arrival rate $\lambda$ is constant, one can reduce $T$ by reducing $\overline{n}$. Thus, scheduling policies like shortest-job-first that try to keep queue size small would improve response times. In a closed system, however, the constant $N$ in $N = XT$ implies a feedback effect: a job that goes through the system quickly immediately returns as another job. Thus, if utilization is high, the scheduling policy essentially just shuffles the jobs in the queue without affecting the average response time. If utilization is low (e.g., the queue is usually empty or has just one job), then scheduling has little impact.

*Principle (vii)* follows from the fact that a partly-open system behaves like an open system when $p \approx 0$, and like a closed system when $p \approx 1$. There is an implicit assumption here that $p$ is independent of load. One can imagine a system (e.g., web surfing [59]) in which $p$ is small (so the user submits few requests) when the system is busy, but $p$ becomes larger when the system is less busy.

Suppose each user in a partly-open system submits $V$ requests to a server with average service time $S$. For user arrival rate $\lambda$, the server utilization is $\rho = \lambda V S$, regardless of think time $Z$. In contrast, a closed queue has $\rho = XVS$, where $Z$ affects the throughput $X$ like in Fig. 6.6. This provides some intuition for *Principle (viii)*.

*DatabaseSerializability* [3]

Serializability is the classical correctness definition for concurrent execution of database transactions. This criterion requires equivalence to a serial execution, so it is too severe for e-commerce: it is unnecessarily stringent, and imposes a costly performance constraint.

This paper proposes a relaxation of the correctness definition, and uses a queueing network to study its impact on transaction performance (whereas *NetworkProcessor* [34] uses a queueing network to explore implementation possibilities). This makes it possible to evaluate the definition even before spending effort to implement the algorithms and use them for a simulation.

The system of master copy, caches, terminals, etc. is schematically illustrated by the models in *Fig. 3* and *Fig. 4*. The major input parameters are listed in *Table 1*. The performance study focuses on throughput, which is the usual metric of interest for online transaction processing.

One can view *Fig. 3* as an example of hierarchical decomposition, where the Ready Queue has a load-dependent server and the restart delay is an $M$-server delay center, as illustrated in Fig. 6.7. The Update Queue and Object Queue in *Fig. 3* are themselves load-dependent queues, with service rates that depend on the resource contention (for CPU and IO). The CC Queue is also load-dependent, with a service rate that depends on data contention (for locks); the underlying concurrency control model also determines the branching probabilities for ACCESS, UPDATE, BLOCK, RESTART, and COMMIT.

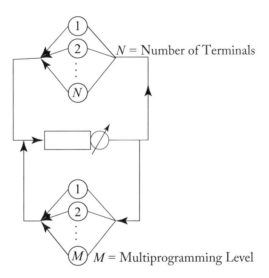

Figure 6.7: A flow-equivalent view of the single-site queueing model in *DatabaseSerializability* [3], where the load-dependent server represents the subnetwork for the Ready Queue.

The MASTER subsystem in *Fig. 4* can also be viewed as a load-dependent queue that represents the subnetwork in *Fig. 3*. This is not the case for the CACHE subsystem, since it includes the Update Queue (Cache $j$), $j = 1, \ldots, n$, in *Fig. 4*. There are thus multiple levels to the hierarchical decomposition.

When a transaction restarts and repeats its requests, it may repeat the data conflicts that caused it to restart. This can be avoided if the restarted transaction waits for the conflicting transactions to finish—hence the RESTART delay in *Fig. 3*. However, if the model resamples the requests for the restarted transaction, then the delay is unnecessary for avoiding conflict repetition. (This delay is omitted from *TransactionalMemory* [24], even though the experiments there do not use resampling.)

This paper's motivation for relaxing the serializability criterion is to trade currency for performance. *Fig. 5* shows this tradeoff in action: when the currency is relaxed from BSAFC-0 to BSAFC-100, the throughput increases from below Strict2PL (a standard correctness criterion) to above it.

The experiments focus on throughput, which is a system-centric view of performance. As the system here is for e-commerce, one should also consider user-centric measures, like response time. Note that, when throughput is nonmonotonic (like in *Fig. 5*) and does not converge on the upper bound (like in Fig. 6.2), then response time also deviates from its lower bound, and can look more like Fig. 2.5.

*Roofline* [66]

This chapter introduces bottleneck analysis in the context of queueing networks, but the concept is widely applicable. Thus, although the shape of the throughput bounds in Fig. 6.2 looks like a roofline, the reason for the sloping line is different: in Fig. 6.2, the line is a bound imposed by Little's Law; on the other hand, the slope for roofline is determined by memory bandwidth.

Nonetheless, a roofline is the minimum of two intersecting lines that represent the memory bandwidth and processor speed, similar to $X \leq \min_i \frac{K_i}{D_i}$ for the queueing network, where each $\frac{K_i}{D_i}$ gives a horizontal line that represents a queue. One can imagine that, in some other context, bottleneck analysis may involve taking the minimum of some other collection of lines and curves.

The system analyzed here has multiple cores and a DRAM (*Table 1*) running a floating-point kernel from the Seven Dwarfs in scientific computing (*Table 2*). Details like multithreading, how threads are distributed over the cores, the cache hierarchy, etc. are not explicitly modeled; "operational intensity" can also have more than one meaning (e.g., exchanges/byte in *Fig. 5* instead of flops/byte). This is similar to the analysis in Sec. 6.1, where details like how the queues are connected, the queueing discipline, and the branching probabilities are not considered. Ignoring such details may give loose bounds on the performance; on the other hand, the bounds are robust with respect to variations and changes in the details.

The issues addressed by the paper are related: (a) Given an architecture and a kernel, is performance constrained by processor speed or memory bandwidth? (b) Given a particular kernel, which architecture will provide the best performance? (c) Given a particular architecture, how can the kernels be optimized (*Table 3*) to push the performance? Here, the performance metric is execution speed (flops or exchanges per second).

The key parameter used to characterize each kernel is operational intensity, i.e., the average number of operations executed per byte of memory traffic between cache and DRAM. A kernel like FFT has an operational intensity that depends on its input.

*PerformanceAssurance* [45]

MAQ-PRO is an approach to resource provisioning for multi-tier services (like *Internet-Services* [63]) at the granularity of software components. It has a Stage I that constructs an analytical model that is used in Stage II (*Fig. 1*) for component replication (to satisfy service level agreements for response time, say) and placement (to minimize resource requirements), as illustrated in *Fig. 14*.

The placement algorithm (*Algorithm 2*) calls on a modified multiclass MVA (*Algorithm 2*) and Schweitzer's approximation to solve the queueing network model (cf. *Fig. 2*) for the system. The modification is necessary for two reasons: First, while standard MVA assumes service demand does not change with queue size, *Fig. 5* shows that this is not so in practice because of context switching, etc. MAQ-PRO therefore uses profiling and regression to derive a piecewise-linear function (*Eqs. (2)* and *(3)*) for how service demand depends on utilization.

Second, although MVA can be extended to multiserver queues to model multiprocessor systems, the extension is computationally expensive. The load-dependent queue is therefore approximated (*Eq. (6)*) by a single-server queue with service rate that has a *correction factor c* which varies with utilization (*Fig. 8*). This $c$ is again obtained empirically by profiling, which avoids the need to model the underlying causes (multiple processors, cache sharing, etc.).

CHAPTER 7

# Deterministic Approximations

There is a common misunderstanding that performance modeling requires queueing theory. This is obviously not the case for networking, say, where queues are pervasive (link layer, routers, etc.), but queueing models are uncommon. Many of the concepts we have introduced via queueing systems (open/closed models, Little's Law, Markov chains, PASTA, bottlenecks, etc.), in fact apply to non-queueing systems. This chapter and the next introduce techniques that involve no queueing theory.

To help make an analysis tractable, random variables are often treated as deterministic. This chapter considers two such techniques: Average Value Approximation and fluid approximation.

## 7.1 AVERAGE VALUE APPROXIMATION (AVA)

Very often in performance modeling, a derivation would make an approximation by replacing a random variable $X$ by its mean $EX$. I call this **Average Value Approximation** (AVA).

To illustrate, although $EXY \neq (EX)(EY)$ in general, AVA gives

$$E(XY) \approx E((EX)Y) = (EX)EY, \quad \text{since } EX \text{ is a constant.}$$

Intuitively, AVA would give a good approximation if $X$ is approximately constant, i.e., $VarX$ is small. For $EXY$, this is certainly true.

**Theorem 7.1**  *Let $X$ and $Y$ be random variables, $EX = \mu_X$, $EY = \mu_Y$ and*

$$\rho_{XY} = \frac{E(X - \mu_X)(Y - \mu_Y)}{\sqrt{(VarX)(VarY)}}.$$

*($\rho_{XY}$ is called the **correlation coefficient**.) Then $|\rho_{XY}| \leq 1$.*  □

It follows that $|EXY - \mu_X\mu_Y| \leq \sqrt{(VarX)(VarY)}$, so $EXY \approx \mu_X\mu_Y$ if $VarX$ is small.

To give another example, although $E(\frac{X}{Y}) \neq \frac{EX}{EY}$ in general, AVA gives

$$E\left(\frac{X}{Y}\right) \approx E\left(\frac{X}{EY}\right) = \frac{1}{EY}EX, \quad \text{since } \frac{1}{EY} \text{ is a constant.}$$

Now recall Renyi's Theorem in Chapter 0. We can use AVA to "derive" the mean in that result by

$$E(X_1 + \cdots + X_N) \approx E(X_1 + \cdots + X_{EN}) \approx (EN)EX_i = \left(\frac{1}{p}\right)m \quad \text{(by Lemma 0.7),}$$

which is the mean for Exponential($\frac{p}{m}$), by Lemma 0.2. Strictly speaking, this argument is incorrect, since $EN$ may not even be an integer—we see here the leap taken by AVA.

AVA is a simplistic approximation that is widely used in analytical modeling (including those surveyed in this book).

## 7.2   FLUID APPROXIMATION

Although jobs are discrete, treating their movement as a continuous fluid flow can significantly simplify the analysis without losing important details. Such an analysis is called a **fluid approximation**. A fluid flow can be described with differential equations, and the following is a review of some linear algebra for stability analysis.

Consider a 2-dimensional real vector $\mathbf{u}(t)$ that is a function of time $t$, and a $2 \times 2$ constant real matrix $A$. For scalar $\psi$ and nonzero $\mathbf{u}$, suppose $A\mathbf{u} = \psi\mathbf{u}$ has two solutions

$$A\mathbf{v}_1 = \psi_1\mathbf{v}_1 \quad \text{and} \quad A\mathbf{v}_2 = \psi_2\mathbf{v}_2.$$

Here, $\psi_1$ and $\psi_2$ are called **eigenvalues**, and $\mathbf{v}_1$ and $\mathbf{v}_2$ are their corresponding **eigenvectors**. If the eigenvectors are linearly independent, then the **homogeneous** differential equation

$$\frac{d\mathbf{u}}{dt} = A\mathbf{u} \quad \text{has general solution } \mathbf{u}(t) = c_1 e^{\psi_1 t}\mathbf{v}_1 + c_2 e^{\psi_2 t}\mathbf{v}_2$$

for some scalars $c_1$ and $c_2$ determined by the initial condition $\mathbf{u}_0 = \mathbf{u}(t_0)$.

The eigenvalues are determined by the **characteristic equation**

$$\det(A - \psi I) = 0, \quad \text{where } I \text{ is the } 2 \times 2 \text{ identity matrix.}$$

For a 2-dimensional $\mathbf{u}$, $\det(A - \psi I)$ is a quadratic in $\psi$. Each root of the characteristic equation is, in general, a complex number $\psi = \beta + i\omega$, for some real numbers $\beta$ and $\omega$, so

$$e^{\psi t} = e^{\beta t}e^{i\omega t} = e^{\beta t}(\cos\omega t + i\sin\omega t).$$

Note that, if a quadratic with real coefficients have roots $\psi_1 = \beta_1 + i\omega_1$ and $\psi_2 = \beta_2 + i\omega_2$, then $\psi_1 + \psi_2$ is a real number, so $\omega_1 = -\omega_2$.

The oscillatory factor $\cos \omega t + i \sin \omega t$ is bounded, and

$$\lim_{t\to\infty} e^{\beta t} = \begin{cases} 0 & \text{if } \beta < 0 \\ \infty & \text{if } \beta > 0 \end{cases}.$$

Hence, the solution

$$\mathbf{u}(t) = c_1 e^{\beta_1 t} (\cos \omega_1 t + i \sin \omega_1 t) \mathbf{v}_1 + c_2 e^{\beta_2 t} (\cos \omega_2 t + i \sin \omega_2 t) \mathbf{v}_2$$

converges to a stable equilibrium at $\mathbf{0}$ (i.e., $\lim_{t\to\infty} \mathbf{u}(t) = \mathbf{0}$) if $\beta_1 < 0$ and $\beta_2 < 0$.

Figure 7.1 shows three possible scenarios for $\mathbf{u}(t)$ to converge from $\mathbf{u}_0$ to a stable equilibrium at $\mathbf{0}$, while Fig. 7.2 shows three possible scenarios for $\mathbf{u}(t)$ to diverge from $\mathbf{u}_0$. For Fig. 7.2, note that $\frac{d\mathbf{u}}{dt} = \mathbf{0}$ at $\mathbf{u} = \mathbf{0}$, so there is an equilibrium at $\mathbf{u} = \mathbf{0}$, but this equilibrium is **unstable** since a slight perturbation will cause $\mathbf{u}$ to diverger from $\mathbf{0}$. There may be no equilibrium for some other combination of eigenvalues, as illustrated in Fig. 7.3.

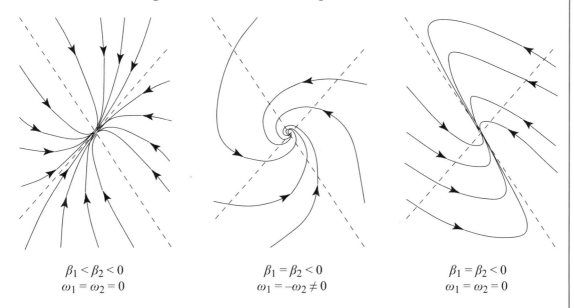

$$\begin{array}{ccc} \beta_1 < \beta_2 < 0 & \beta_1 = \beta_2 < 0 & \beta_1 = \beta_2 < 0 \\ \omega_1 = \omega_2 = 0 & \omega_1 = -\omega_2 \neq 0 & \omega_1 = \omega_2 = 0 \end{array}$$

Figure 7.1: Convergence from $\mathbf{u}_0$ to a stable equilibrium at $\mathbf{0}$, and corresponding eigenvalues.

Stability can also be characterized by the trace and determinant. If $A = \begin{pmatrix} a & b \\ c & d \end{pmatrix}$, then

$$\det(A - \psi I) = (a - \psi)(d - \psi) - bc = \psi^2 - \text{trace}(A)\psi + \det(A) = 0,$$

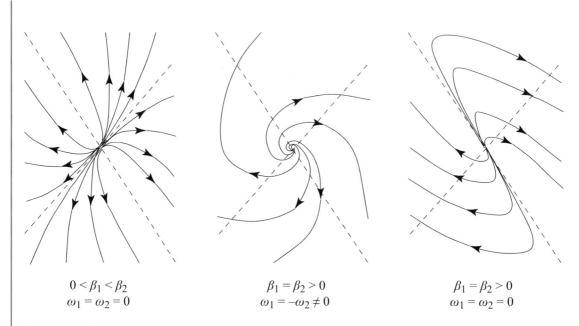

Figure 7.2: Divergence from $\mathbf{u}_0$, and corresponding eigenvalues.

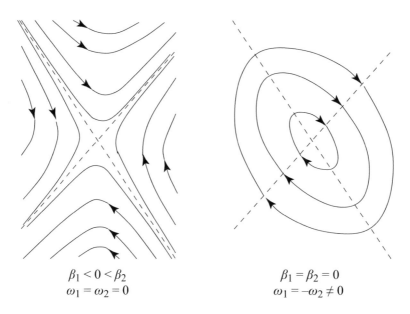

Figure 7.3: Other possible trajectories and corresponding eigenvalues.

so the roots are

$$\psi_1 = \frac{\text{trace}(A) + \sqrt{\text{trace}(A)^2 - 4\det(A)}}{2} \quad \text{and}$$

$$\psi_2 = \frac{\text{trace}(A) - \sqrt{\text{trace}(A)^2 - 4\det(A)}}{2}.$$

**Case** $\text{trace}(A) < 0$ and $\det(A) > 0$:

If $\text{trace}(A)^2 - 4\det(A) < 0$, then $\omega_1 \neq 0$, $\omega_2 \neq 0$ and $\beta_1 = \beta_2 = \text{trace}(A)/2 < 0$. If $\text{trace}(A)^2 - 4\det(A) \geq 0$, then $\omega_1 = \omega_2 = 0$ and $\text{trace}(A) + \sqrt{\text{trace}(A)^2 - 4\det(A)} < 0$ (since $\det(A) > 0$), so $\beta_1 < 0$ and $\beta_2 < 0$. Either way, $\mathbf{u}(t)$ has a trajectory that converges to a stable equilibrium at $\mathbf{0}$, as illustrated in Fig. 7.1.

**Case** $\text{trace}(A) > 0$ and $\det(A) > 0$:

If $\text{trace}(A)^2 - 4\det(A) < 0$, then $\omega_1 \neq 0$, $\omega_2 \neq 0$ and $\beta_1 = \beta_2 = \text{trace}(A)/2 > 0$. If $\text{trace}(A)^2 - 4\det(A) \geq 0$, then $\omega_1 = \omega_2 = 0$ and $\text{trace}(A) - \sqrt{\text{trace}(A)^2 - 4\det(A)} > 0$ (since $\det(A) > 0$), so $\beta_1 > 0$ and $\beta_2 > 0$. Either way, $\mathbf{u}(t)$ diverges from $\mathbf{0}$, as illustrated in Fig. 7.2.

For an **inhomogeneous system**

$$\frac{d\mathbf{u}}{dt} = A\mathbf{u} + \mathbf{b}, \quad \text{where } A \text{ is invertible and } \mathbf{b} \text{ is constant,}$$

the steady state (particular) solution is $\mathbf{u} = -A^{-1}\mathbf{b}$, so the general solution is

$$\mathbf{u}(t) = c_1 e^{\beta_1 t}(\cos\omega_1 t + i\sin\omega_1 t)\mathbf{v}_1 + c_2 e^{\beta_2 t}(\cos\omega_2 t + i\sin\omega_2 t)\mathbf{v}_2 - A^{-1}\mathbf{b}.$$

For the eigenvalues in Fig. 7.1, the trajectories are translated, and converge at $-A^{-1}\mathbf{b}$.

## 7.3   DISCUSSION OF PAPERS

AVA and fluid approximations were used in some of the papers we have discussed so far, as well as in another five (*TCP* [43], *BitTorrent* [44], *NoC* [29], *Gossip* [2], and *CachingSystems* [36]) that we introduce below.

*StreamJoins* [28]

In *Table 1*, the number of tuples $B$ is a random variable for a logical window, and the number of hash buckets $|B|$ is also not a constant. One can therefore view $B/|B|$ in *Eq. (8)* as an example of AVA.

*GPRS* [42]

The expression $\tilde{X} = \frac{\tilde{U}}{\tilde{Q}}$ in *Eq. (10)* is a similar example of AVA.

*StorageAvailability* [18]

The authors use the machine repairman model to derive *Eq. (1)*, but the formula suggests that there is a combinatorial derivation. In fact, one can use the following AVA argument: The failure rate is $\lambda$ and the recovery rate is $\mu$, so the average up-time between consecutive failures is $1/\lambda$, and the average down-time for recovery is $1/\mu$. By AVA, the availability for an SE (i.e., the probability that it is up) is

$$\frac{\frac{1}{\lambda}}{\frac{1}{\lambda} + \frac{1}{\mu}} = \frac{1}{1 + \rho}.$$

It is clear that this solution holds even if $\lambda \geq \mu$ (so $\rho \geq 1$). Since the SEs are assumed to be independent, the probability that at least $Q$ SEs are available is binomial, i.e.,

$$\sum_{j=Q}^{N} \binom{N}{j} \left(\frac{1}{1+\rho}\right)^j \left(\frac{\rho}{1+\rho}\right)^{N-j},$$

which is equivalent to *Eq. (1)*. This argument does not assume the distributions are exponential, and may be more readily accepted by engineers. However, it still assumes independence of failures.

*TransactionalMemory* [24]

The use of $L - \ell_w E(Q)$, instead of $L - \ell_w i$ in the denominators for *Eq. (3)* is an example of AVA. *Eq. (12)* has $E(Q)$ on both sides, so the model results in a fixed-point approximation that is solved iteratively.

*SensorNet* [49]

One can derive *Eq. (3)* with the following AVA argument: Given a random walk on a torus, with all directions equilikely, a MULE is equally likely to be anywhere along its path. If the expected path length is $E[R_i]$, then the probability of being at grid point $i$ is $\pi_i = 1/E[R_i]$.

One can do a similar derivation of *Corollary 2.1*: By *Eq. (3)* the average inter-arrival time of a specific MULE to a sensor is $N$, so the arrival rate is $1/N$; since there are $N_{mules}$, the arrival rate of MULEs to that sensor is $N_{mules}/N = \rho_{mules}$; the inter-arrival time of MULEs is therefore $1/\rho_{mules}$, which gives the buffer occupancy in *Eq. (17)*.

*RouterBuffer* [1]

Since the burst size is $X$, the expression for $E[N]$ estimates the queue size in terms of bursts, and is thus larger than the packet queue size estimated by $E[Q]$. The authors correct the difference by estimating half a burst as remaining whenever there is a busy server, so the correction term is Prob(server busy)$\frac{EX}{2} = \rho\frac{EX}{2}$. This is an example of AVA.

*SoftErrors* [40]

The AVF model estimates a critical path as taking $\ell K(W)$ cycles, where $\ell$ is average instruction latency $\ell$, and $K(W)$ is the average critical path length for an instruction window of size $W$; this is an example of AVA.

*ServerEnergy* [19]

ACES models reliability cost by the cost per duty (boot) cycle. Since the number of duty cycles is a random variable, the model uses AVA and divides total cost by the *average* number of duty cycles MTTF (mean time to failure).

*TCP* [43]

The system in this paper refers to a bulk transfer in a TCP Reno connection over multiple routers. The central issue is how packet loss and timeout affect the throughput via the protocol (window adjustment, retransmission, etc.)—see *Eq. (31)*. The model uses the technique of discretizing time into *rounds* (*Fig. 2*).

Instead of making various assumptions (that are hard to verify or justify, considering the many approximations), one could just view some of the derivations as an application of AVA. For example, instead of considering $\{W_i\}_i$ to be a Markov regenerative process, we could just use AVA to get $B = \frac{EY}{EA}$ in *Eq. (1)*. Similarly, one could forego the assumption of mutual independence between $\{X_i\}$ and $\{W_i\}$ and derive *Eq. (12)* from *Eq. (10)* by AVA.

Rather than assume round trip times to be independent of congestion window size and round number, one could use AVA to derive *Eq. (6)*:

$$EA_i = E \sum_{j=1}^{X_i+1} r_{ij} \approx E \sum_{j=1}^{EX_i+1} r_{ij} = \sum_{j=1}^{EX+1} Er_{ij} = (EX+1)Er,$$

where the approximation regards $EX$ to be an integer.

As the formulas become more complicated, it is no longer clear what assumptions are sufficient for the derivations, and the authors simply apply AVA in *Eq. (25)*:

$$Q = E\hat{Q}(W) \approx \hat{Q}(EW).$$

*Eq. (19)* is an example of a fixed-point approximation (Sec. 5.3): TCP throughput $B(p)$ is expressed in terms of loss probability $p$, which in turn depends to $B(p)$.

*BitTorrent* [44]

The system consists of peers sharing files through a chunk protocol that is similar to Bit-Torrent; a peer becomes a seed when it has finished downloading the desired file. Notice that the centralized tracker, while crucial to the functioning of the system, is omitted from the performance model.

The move from a server-client system to one that is peer-to-peer is basically motivated by scalability, and the model is used to study this and related issues—the growth in number of seeds, the impact of download abortion and bandwidth constraint, the effectiveness of the incentive and unchoking mechanisms.

The main parameters for the model are the request arrival rate $\lambda$, the uploading bandwidth $\mu$, the downloading bandwidth $c$, the abort rate $\theta$ and the peer departure rate $\gamma$.

The model is a deterministic fluid approximation that describes the evolution in number of downloaders $x(t)$ and seeds $y(t)$ as a pair of ordinary differential equations. In *Eq. (1)*, $\lambda - \theta x$ corresponds to the transition rate from state $\langle x, y \rangle$ to $\langle x + 1, y \rangle$ in a Markov chain model, $-\gamma y$ the transition rate from $\langle x, y \rangle$ to $\langle x, y - 1 \rangle$, and $\min(cx, \mu(\eta x + y))$ the transition rate from $\langle x, y \rangle$ to $\langle x - 1, y + 1 \rangle$. Little's Law is used to derive the important performance measure $T$ in *Eq. (6)*.

After deriving the steady state, its stability is analyzed by examining the eigenvalues of the pair of equations. Note that the $\lambda$ term makes the pair inhomogeneous.

The model here is an interesting contrast to that in *MediaStreaming* [62]: although both consider peer-to-peer downloading, one studies steady state solution with a fluid approximation, while the other is a transient analysis using discretized time.

## *NoC* [29]

On-chip interconnects are replacing buses and point-to-point wires. The performance, fault-tolerance, and energy issues for such network-on-chip are inter-related. For example, errors from crosstalk, electromagnetic interference and cosmic radiation (of increasing importance with miniaturization) can cause packet retransmissions, thus increasing latency and energy consumption. This paper proposes a queueing model as a design tool for fast evaluation of router designs.

*Fig. 1* illustrates the system under study. Major input parameters to the queueing model are number of hops $H$, number of flits per packet $M$, injection probability $P_{pe}$, and flit buffer size $D$. The performance model focuses on latency $T$ (*Eq. (3)*) and the power metric $P_R$ (*Eq. (25)*).

Wormhole routing on an NoC is like a virtual circuit in a wide-area network: the flits of a packet follow the header flit's path like a worm, simultaneously occupying multiple nodes. This means the architecture cannot be modeled as a queueing network (since a job in a queueing network does not occupy multiple queues).

The authors discretize time by analyzing what happens in each cycle. This also allows them to switch from considering an arrival rate to deriving an event probability in a cycle, in the following sense: Suppose the arrival rate at an input virtual queue is $\lambda$ flits per cycle; since at most 1 flit can be processed per cycle at any queue, we have $\lambda < 1$, so $\lambda$ can be interpreted as the probability that there is a flit in that cycle. This is similar to utilization being also a probability in Eq. (2.3) for a one-server queue. This relationship induced by Little's Law between arrival rate and probability can also be seen in *Eq. (6)*.

Similarly, the flit injection probability $P_{pe}$ is the arrival rate of flits per cycle. If there are $R$ routers, then the total flit arrival rate at all queues is $RP_{pe}H$, and this is divided among the $RN$ queues to give $P_c$. One must take care to apply sanity checks when making approximations in a model, and we see here that *Eq. (4)* puts a limit on $H$ (since $P_c < 1$). *Eq. (7)* also requires $P_h < 1 - P_{con\_va}$.

The Markov chain in *Fig. 3* is for an $M/M/1/D$ queue. However, the authors use the solution from $M/M/1$ for $r = 0, 1, 2, \ldots$ to approximate the blocking probability in *Eq. (17)*. *Eq. (19)* shows the use of the $M/M/1$ expression for average waiting time (see Eq. (2.8)). Incidentally, the "service time" is obviously not memoryless, since flits have bounded size.

The model requires an iterative solution because there are cyclic dependencies among the variables. For example, $P_{ready\_busy}$ depends on $P'_h$ (*Eq. (6)*), $P'_h$ depends on $P_{con\_va}$ (*Eq. (7)*), and $P_{con\_va}$ depends on $P_{ready\_busy}$ (*Eq. (10)*).

### Gossip [2]

In a mobile network, data can be propagated to nodes via a gossip protocol that opportunistically exchange data via wireless links during transient encounters. Arguably, dissemination by gossip is also applicable if the nodes are stationary but joins and leaves the network. However, there is a crucial difference: a node that leaves may remove the only copy of an item.

The main issue analyzed by the model is the optimal value for $s$, the number of items to exchange between two nodes if the network is stationary and static (*Sec. 5*). Intuitively, if $s$ is too small, a cache takes too long to acquire new items; on the other hand, if $s$ is too big, a cache may be forced to discard newly acquired items. The metric to optimize is $x(t)$, the fraction of nodes that carry a copy of an item at time $t$, where $x(0) = 1/N$ and $N$ is the number of nodes. The other input parameters are the cache size $c$ in each node, and the total number of data items $n$.

*Eq. (1)* for $P_{drop}$ is derived by a combinatorial argument. Although it explicitly expresses $P_{drop}$ in terms of $n$, $c$, and $s$, it is in a form that makes further analysis difficult. In contrast, AVA yields the approximation $\frac{n-c}{n-s}$ for $P_{drop}$, thus giving closed-form expressions for the probabilities $P(11|10)$, $P(10|11)$, etc. and facilitating the analysis in *Sec. 5*. Note also the use of a fluid approximation to analyze the round-based (i.e., discrete time) protocol.

### Caching Systems [36]

A basic caching system has a limited amount of cache for holding objects (retrieved from slower memory), a stream of requests for objects, and some policy for inserting and re-placing objects in the cache. The key performance metric is $p_{\text{hit}}$, the probability that a requested object is in the cache.

Analyzing how the interaction between the request pattern and the replacement policy can be very difficult, so most analyses are based on a simplified pattern—like the independent reference model (IRM)—or an idealized policy (e.g., pure LRU, no prefetching, etc.). This paper offers a model that goes beyond IRM and covers multiple policies.

The model starts with the approximation in *Eq. (1)* for IRM plus LRU. This approximation *decouples* the request streams, so that $p_{\text{hit}}$ for object $m$ only depends on request rate $\lambda_m$. This decoupling is unintuitive since, given a limited cache size $C$, an object $m$ may be evicted because of requests for other objects. The impact of $C$ is, in fact, indirectly reflected in the *cache eviction time* $T_C(m)$, and the decoupling assumes $C$ is large.

Since $T_C(m)$ is the time for $C$ distinct requests different from $m$, it is a random variable. The analysis however treats $T_C(m)$ as a constant, so this is an example of AVA. In fact, the derivation further considers $T_C(m)$ as a constant $T_C$ that is the same for all $m$. In *Sec.IV-C*, however, $T_C$ is again treated as a random variable (but independent of $m$), so Erlang's formula for blocking probability can be applied to both *RANDOM* and *FIFO*.

The $M/G/1/1$ formula assumes Poisson arrivals. This happens under IRM because the inter-request time for object $m$ is a geometrically distributed sum of identically distributed random variables, so it is exponentially distrbuted (approximately, by Theorem 0.10). The paper thus uses queueing theory to get *Proposition 1* for different replacement policies, even though there does not appear to be a queue for any object.

The paper relaxes the reference pattern from the exponential distribution to a general $F_R(m,t)$ and uses, as an example, a hyperexponential distribution with two branches (Fig. 2) and rates $z\lambda_m$ and $\lambda_m/z$. A larger $z$ implies a smaller mean inter-request time and, in that sense, increases temporal locality. However, this does not model other temporal correlation; e.g., the inter-request times for an object referenced in a loop, or for two objects in the same data structure. Spatial locality (a sequential scan, say) is also not covered by $F_R(m,t)$.

CHAPTER 8

# Transient Analysis

Analyzing transient behavior is generally a difficult task; for example, Little's Law no longer applies. This is why, except for the stability analysis in the previous chapter, we have focused on the steady state.

In this chapter, we apply decomposition to do a different stability analysis of the equilibrium for an interactive system. We then use a fluid approximation to study the spread of a worm over the Internet—this is an example where interest lies in the transient, not the steady state.

## 8.1    DECOMPOSING AN EQUILIBRIUM

Consider a closed interactive system, with $N$ users, $K$ of whom are waiting for jobs to finish, and average think time $Z$. Using flow equivalence:

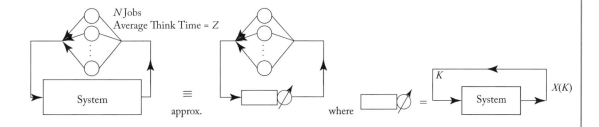

By Little's Law, $N = K + X(K)Z$, where $X(K)$ is the throughput. We can rewrite this as $\lambda_{in} = \lambda_{out}$, where $\lambda_{in} = \frac{N-K}{Z}$ and $\lambda_{out} = X(K)$, and interpret $K$ as determined by an equilibrium between an inflow $\lambda_{in}$ and an outflow $\lambda_{out}$, as in Fig. 8.1.

For fixed $N$ and $Z$, $\lambda_{in}$ is a decreasing linear function of $K$. If $X(K)$ is monotonic increasing, we get an intersection between $\lambda_{in}$ and $\lambda_{out}$, thus determining a unique equilibrium point, as illustrated in Fig. 8.2a.

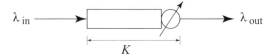

Figure 8.1: The equilibrium in $K$ viewed as a balance between a demand inflow and a supply outflow.

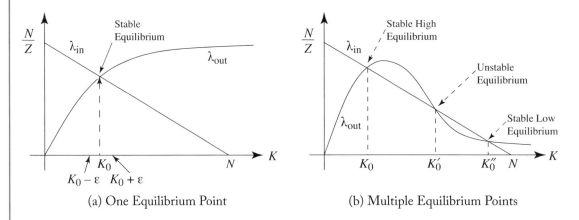

(a) One Equilibrium Point          (b) Multiple Equilibrium Points

Figure 8.2: (a) A stable equilibrium at $K_0$ is a balance between demand $\lambda_{\text{in}}$ and supply $\lambda_{\text{out}}$. (b) If $X(K)$ eventually decreases, there may be three equilibrium points: one high and stable, one unstable, one low and stable [9].

We can view $\lambda_{\text{in}}$ as **work demand** generated at rate $1/Z$ by $N - K$ users, $\lambda_{\text{out}}$ as **work supply** provided by the system to $K$ jobs, and the equilibrium as a balance between demand and supply.

If we adopt a fluid approximation to view the job flow in and out of the queue in Fig. 8.1, we can get an intuitive way of examining the stability of the steady state in Fig. 8.2a, as follows: Suppose the equilibrium is at $K = K_0$. Consider a small $\varepsilon$, $\varepsilon > 0$. At $K = K_0 - \varepsilon$, we have $\lambda_{\text{in}} > \lambda_{\text{out}}$, so $K$ increases (see Fig. 8.1) and moves toward $K_0$. Similarly, at $K = K_0 + \varepsilon$, we get $\lambda_{\text{in}} < \lambda_{\text{out}}$, so $K$ decreases and, again, moves toward $K_0$. Thus, a small transient fluctuation from $K_0$ will cause $K$ to return to $K_0$, so the equilibrium is **stable**.

In real systems, however, a large $K$ often results in inefficiencies (overheads, resource contention, etc.), that cause $X(K)$ (i.e., $\lambda_{\text{out}}$) to eventually decrease. For such a system, we can get multiple equilibrium points, as illustrated in Fig. 8.2b.

However, their stability may be different. In Fig. 8.2b, the equilibrium points at $K_0$ and $K_0''$ are stable, but the one at $K_0'$ is not: a small transient fluctuation in $K$ from $K_0'$ will cause $K$

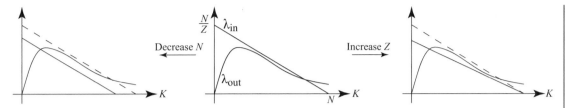

Figure 8.3: A system can return to a high equilibrium if $N$ decreases or $Z$ increases.

to move further from $K_0'$, driving it to a stable equilibrium at $K_0$ or $K_0''$. Since the equilibrium at $K_0'$ is unstable, it is impossible to observe in a real system, or even in simulations. Only an analytical model can reveal this $K_0'$, this boundary beyond which the system would slip further out to a stable but low equilibrium.

A stable but low equilibrium (like $K_0''$) is undesirable. How can a system get out of such a degraded state? Intuitively, performance should improve if $N$ is reduced (by admission control, say). We can see this in Fig. 8.3, where reducing $N$ causes a parallel shift in the $\lambda_{\text{in}}$ line, possibly relocating the equilibrium to a point where performance is higher. The system can also recover by itself through increasing $Z$ (at night, say). This causes the $\lambda_{\text{in}}$ line to tilt and, again, relocate the equilibrium.

Here, we see how the simple idea of decomposing the system equilibrium into $\lambda_{\text{in}}$ and $\lambda_{\text{out}}$ helps us understand its dynamics.

## 8.2    EPIDEMIC MODELS

We now do a transient analysis of how an infection spreads among hosts. The system consists of hosts that are linked to each other; if some hosts are **infected** (by a worm, say), the infection can spread like an epidemic to others through the links.

Let $y$ be the fraction of hosts in the system that are infected. The following is the differential equation for a classical epidemic model:

$$\frac{dy}{dt} = ky(1-y) \quad \text{for } 0 < y < 1, \quad \text{where } k > 0 \text{ and } y_0 = \lim_{t \to 0} y. \tag{8.1}$$

Let the number of hosts be $N$. $N$ is an integer, so $y$ is—strictly speaking—not a continuous variable. In treating it as one, we are thus using a fluid approximation.

Now, $ky(1-y) = k(yN)(N-yN)/N^2$, where $yN$ and $N-yN$ are the number of infected and uninfected hosts. If any infected host can infect any uninfected host, then Eq. (8.1) models the infection rate as proportional to the number of possible infections. After rearranging, integration gives

$$\int \left(\frac{1}{y} + \frac{1}{1-y}\right) dy = k \int dt$$

so $\quad \ln y - \ln(1-y) = kt + \text{constant}$

i.e., $\quad y = \dfrac{1}{1 + \left(\frac{1}{y_0} - 1\right)e^{-kt}}.$

For small $t$, we have $e^{-kt} \approx 1$, so

$$\frac{1}{1 + \left(\frac{1}{y_0} - 1\right)e^{-kt}} = \frac{1}{1 - e^{-kt} + \frac{1}{y_0 e^{kt}}} \approx y_0 e^{kt},$$

i.e., $y$ starts out by increasing exponentially. Also, $\frac{dy}{dt} > 0$ for $0 < y < 1$, so

$$\frac{d^2 y}{dt^2} = k\frac{dy}{dt}(1 - 2y) \begin{cases} > 0 & \text{for } y < \frac{1}{2} \\[2mm] < 0 & \text{for } y > \frac{1}{2} \end{cases}.$$

Therefore, there is an inflection point when $y = \frac{1}{2}$, marking a change of phase in the epidemic.

The model above assumes that an infected host is forever infectious. If, instead, an infected host can be **removed** (i.e., it is no longer infectious because it recovers or dies), then one might adopt the **Kermack-McKendrick Model**: Let $N$ be the number of hosts, $S$ the number of hosts that are susceptible to infection, $I$ the number of hosts that are (infected and) infectious, and $R$ the number of hosts that are (infected but) removed; then

$$\frac{dS}{dt} = -\beta S I \quad \text{where } \beta > 0 \tag{8.2}$$

$$\frac{dR}{dt} = \gamma I \quad \text{where } \gamma > 0 \tag{8.3}$$

$$N = S + I + R. \tag{8.4}$$

Here, $\beta$ is the **infection rate** per susceptible-infected pair, and $\gamma$ is the **recovery rate** per infected host.

Equations (8.2) and (8.3) give

$$\frac{dS}{dR} = -\frac{1}{\rho}S \quad \text{where } \rho = \frac{\gamma}{\beta},$$

with solution $S = S_0 e^{-\frac{1}{\rho}R}$, so $S$ decays exponentially with $R$. Next, Eq. (8.4) gives

$$\frac{dI}{dt} = -\frac{dS}{dt} - \frac{dR}{dt} = -\frac{dS}{dR}\frac{dR}{dt} - \frac{dR}{dt} = \frac{1}{\rho}\frac{dR}{dt}(S - \rho).$$

Therefore, for $I \neq 0$ (see Eq. (8.3)),

$$\frac{dI}{dt} > 0 \quad \Longleftrightarrow \quad S > \rho;$$

this characterizes the time at which infection begins to decrease. In fact, this also says that if $S < \rho$ initially, then $dI/dt < 0$ and there will be no epidemic. Note that we can draw this conclusion from the model even though we do not have a closed-form solution, thus illustrating the power of an analytical model (see Sec. 10.1.1).

## 8.3    DISCUSSION OF PAPERS

Below, we discuss the equilibrium and transient behavior in some of the previous papers, then introduce four papers (*CodeRed* [69], *DistributedProtocols* [20], *EpidemicRouting* [67], and *InformationDiffusion* [37]) that model systems with differential equations.

*InternetServices* [63]

    The paper adopted a closed interactive model ($Z = 1$ sec), so its equilibrium can be analyzed by a decomposition into $\lambda_{\text{in}}$ and $\lambda_{\text{out}}$. A separable queueing network cannot model efficiency loss at high load, so $\lambda_{\text{out}}$ will be monotonic increasing (unlike Fig. 8.3) and there will be a single, stable equilibrium.

    The CPU bottleneck (*Figs. 7* and *8*) imposes a bound on $\lambda_{\text{out}}$, which therefore flattens out as the number of sessions $N$ increases. With a similar plateau for response time (*Figs. 4, 8*, and *9*), Little's Law says that the number of concurrent sessions in the three tiers also becomes constant. This means that most of the increase in $N$ is actually in the think stage $Q_0$ and $Q_i^{\text{drop}}$ at the bottleneck in *Fig. 5*.

*MediaStreaming* [62]

    The determination of $k_0$, when there are enough peers to take over the servers' role in satisfying requests, requires a transient analysis. By discretizing time and writing a difference equation for $P(k)$, i.e., *Eq. (1)*, the authors demonstrate an alternative to modeling with $dP(t)/dt$.

*DatabaseSerializability* [3]

    If we apply demand-supply equilibrium decomposition to the model, the throughput becomes the supply curve $\lambda_{\text{out}}$. From *Fig. 5*, we see that with a relaxed time constraint, BSAFC-100 has throughput looking like $\lambda_{\text{out}}$ in Fig. 8.2a. With a tighter time constraint, however, BSAFC-10 has throughput that looks like $\lambda_{\text{out}}$ in Fig. 8.2b, so there may be multiple equilibrium points. Whether one of these will result in a stable low equilibrium depends on how low the tail gets for the throughput curve.

*DependabilitySecurity* [61]

Fig. 4 uses reliability and availability to classify dependability. **Availability** is usually defined as the steady state probability that a system is working. **Reliability**, however, is usually defined as the probability that the system is working up to time $t$; this probability is determined by some transient analysis.

For example, the reliability model in *Fig. 2* is a Markov chain for a workstation file system server system; if $T$ is the time it takes for the system to go from the initial state ($\langle 2, 1 \rangle$ for 2 workstations and 1 file server in working order) to a failed state ($\langle 0, 1 \rangle$, $\langle 1, 0 \rangle$, or $\langle 2, 0 \rangle$), then the reliability at time $t$ is $\text{Prob}(T > t)$. The safety model of *Fig. 21* is a similar reliability model.

Like the performance and availability model in *Fig. 14*, *Fig. 13* uses a Markov chain that models a sensor network's confidentiality in one dimension and failure in another dimension. A crucial difference is that this is a transient model with an **absorbing** failure state $F$.

In contrast, the failure state $F$ in *Fig. 16* is not an absorbing state—restoration or reconfiguration can bring the system back into a good state $G$, so one can calculate its steady state availability. In this case, the model is a semi-Markov process, which is a generalization of Markov chains.

In a Markov chain, the system stays in a state for an exponentially-distributed length of time. This assumption may be too strong. For example, some states in *Fig. 16* for an intrusion-tolerant system model attacker behavior, which one should not assume is memoryless. If the state residence time is not memoryless, we get a **semi-Markov process**. The $M/G/1$ queue (Sec. 2.3) is an example of such a process, and *Fig. 19* shows another semi-Markov model for system maintainability, where the transition rates vary with time.

A steady state model can be perturbed to give a transient analysis. For example, the authors briefly describe how survivability for a telecommunication system can be calculated with the Markov chain in *Fig. 18* by forcing a failure, i.e., initializing $\text{Prob}(\langle i, j \rangle)$ to 0, where $\langle i, j \rangle$ is the state for $i$ channels and $j$ calls.

Conversely, *Fig. 15* shows how the transient susceptible-infected-removed model in *Sec. 8.2* can be re-interpreted and extended into a Markov chain for modeling the steady state behavior of an intrusion-tolerant database system. Note that adding a new state $M$ (Malicious) to a Markov chain is much easier than adding a new variable to a set of differential equations.

*PipelineParallelism* [41]

*Sec. 3.3* models work stealing with one queue per thread. With queues stealing tasks from each other, the assumptions underlying separability are broken, so Jackson's Theorem does not apply. Instead, the paper adopts the idea [57] of introducing virtual servers (*Fig. 8*). We see here that, although work stealing happens only when there is transient imbalance in workload, its impact can be modeled with a steady state solution.

*Gossip* [2]

The fluid approximation provides a transient analysis, resulting in *Eq. (4)* that describes how the fraction $x(t)$ of nodes holding an item evolves. In fact, *Eq. (3)* is an epidemic model like Eq. (8.1), and the solution is similar.

*SoftErrors* [40]

At a macro level, the contents of a microarchitectural structure (like reorder buffer) appear dynamic. However, by using disruptive events (e.g., cache miss, mispredictions) to divide the timeline, the model can now treat each interval as in steady state and apply Little's Law. The events themselves require a separate transient analysis on how occupancy changes; a key contribution of this paper lies in analyzing the interaction of two events (*Fig. 3a*).

This technique of **interval analysis** [15] enables a model to factor in fluctuations from one steady state to another. However, it requires some knowledge of the nature and timing of these fluctuations; in the case of *SoftErrors* [40], this knowledge is extracted by the profiler.

*ProactiveReplication* [13]

The idea (in *Sec. 4.1*) of dividing time into intervals is similar to that in *SoftErrors* [40], but there are no special events (like cache misses) here for this discretization. Using intervals of varying length $\Delta T$ determined by a fixed number $D$ of disconnections helps $\Delta T$ adapt to variation in disconnection rate, but it is not clear why an interval chosen this way would be in steady state (an assumption required by Little's Law).

*Fig. 2* is another example of a semi-Markov process, like the one in *DependabilitySecurity* [61]. It is inherently transient, since every peer eventually disconnects permanently. However, the system (*Fig. 3*) can have a steady state as long as peers join and fragments are replicated.

*CodeRed* [69]

The system in this paper refers to a worm spreading among Internet hosts through TCP connections. The main parameters are the number of hosts $N$, the infection rate $\beta$ and the removal rate $\gamma$. The central performance metric is the number of infected hosts $I$.

The basic issue is how $I$ varies with time as infection spreads. The paper seeks to craft a model that matches the measured growth and decay in $I$ for the Code Red worm, and use that to analyze the impact of countermeasures, congestion and topology. This is necessarily a transient analysis; in the steady state, all hosts would have become infected or immune, so $dI/dt = 0$.

The massive number of Internet hosts involved in a worm propagation makes it reasonable to adopt a deterministic model using differential equations. The paper started with the classical epidemic model, with *Fig. 4* illustrating the initial exponential increase in $y$ and the phase change at $y = \frac{1}{2}$. Although these features match the growth in *Fig. 1* and *Fig. 2*, the dying phase of the worm requires a model for the countermeasures. For this, the authors adopted the Kermack-McKendrick model. The $S > \rho$ condition in *Eq. (6)* then locates the turning point in *Fig. 1*.

*DistributedProtocols* [20]

This paper proposes a framework for designing distributed protocols through differential equations; this way, one can leverage on the latter's rich theory to derive desirable properties (stability, scalability, etc.) for the protocols.

The system is a set of processes connected by an asynchronous network. The input parameters are the number of processes $N$ and the constants ($\alpha$, $\beta$, etc.), in the differential equations, and the metrics are the variables ($x$, $y$, etc.). The model is simply the equations themselves.

The equations are a fluid approximation, treating the variables as continuous even though they represent processes (and so are discrete). Moreover, changes in process state happen at the beginning of protocol periods and are prompted by the exchange of messages, so system evolution is also discrete.

The framework transforms the differential equations for $N$ processes into a state machine (like *Fig. 1*) for 1 process: a variable becomes a state, and a term becomes an action. The state machine can, in turn, be interpreted as a Markov chain, but with transition rates that depend on the states—e.g., the transition rate from **receptive** to **stash** in *Fig. 1* is $\beta y$, where $y = \text{Prob}(\textbf{stash})$. Indeed, the balance equations are the original differential equations at equilibrium.

(This interpretation of a fluid model as a Markov chain has a converse that we have already seen: Recall the modification of the Markov chain in *TransactionalMemory* [24] described in Chapter 3, where there is a transition from committed state $k + 1$ to start state 0, so the model becomes the flow diagram for the state of the entire system. Each node $j$ in the diagram then represents all transactions holding $j$ locks, and the edges represent the flow of transactions. In other words, the Markov chain becomes a fluid approximation.)

Naively, one might model the system of $N$ processes as an $N$-dimensional Markov chain, with state space $\{\langle s_1, \ldots, s_N \rangle \mid s_i \in \{\textbf{receptive}, \textbf{stash}, \textbf{averse}\}\}$. An aggregation of these $3^N$ states may cause an unnecessary loss of information. Instead, the approach here is to solve the per-process Markov chain (i.e., *Fig. 1*) in each dimension. If desired, one could construct the $N$-dimensional chain by taking the product of these 1D chains.

The stability analysis we do in Fig. 8.2 is similar to the perturbation analysis, where Gupta started the system at $\langle x_\infty(1 + u), y_\infty(1 + v), z_\infty(1 + w) \rangle$ for some small $u$, $v$ and $w$. By introducing the new variable $t = \dot{u}$, the perturbed equations reduce to a homogeneous system (*Eq. (4)*); one can then examine the stability of the system through the eigenvalues (Sec. 7.2).

*EpidemicRouting* [67]

Unlike *Gossip* [2], where the goal is data dissemination, the protocol here aims to route a packet from a source to its destination via forwarding when two mobile nodes meet wirelessly. This forwarding is akin to infection except, being a protocol, there are multiple variations to the infection mechanism. One objective of this paper is to propose a uniform framework, using differential equations, for modeling the variations.

Although the nodes move continuously, the system state makes a discrete transition each time a packet is forwarded. A Markovian model of these transitions is possible, but its solution is computationally intractable, and extracting analytical insight is difficult. Instead, a fluid approximation of the transitions give differential equations that facilitate numerical evaluation and provide closed-form expressions.

As in the case of MVA (Chapter 5), with this reduction in computational complexity, one gets only the moments and not the distribution of the random variables.

The system consists of wireless nodes moving in some pattern over a restricted area. Each node is a source and a destination for a flow of packets, has a buffer to hold packets, and runs the routing protocol. The main issue is how long it takes for a packet to be delivered to its destination, and the cost in terms of number of copies transmitted, and the buffer space necessary for holding them.

The metrics of interest are therefore the packet delivery delay $T_d$, the number of copies $C_{ep}$ of a packet at delivery time, and the buffer space $Q_{total}$ occupied by all copies of all packets.

There are $N + 1$ nodes, a packet flow has rate $\lambda$ (different from $\lambda = N\beta$ defined for *Eq. (2)*) and, for probabilistic forwarding, a received packet is accepted with probability $p$. Other parameters that specify the roaming area, transmission range and mobility pattern are indirectly specified through $\beta$, the rate at which one node meets another (*Eq. (1)*).

The analysis starts with the classical epidemic model in *Eq. (2)* and a variant of the Kermack-McKendrick model in *Sec. 2.3*, before refinements to suit different forwarding schemes (*Sec. 3*). The authors estimate the number of copies of a packet during its lifetime by $\int_0^\infty I(t)\,dt/L$—this is an example of AVA.

Modeling the finiteness of the buffer requires more variables and equations. Closed-form solutions are no longer possible, and *Sec. 5* uses a fixed-point approximation to solve the equations.

*InformationDiffusion* [37]

Information diffusion (of memes, rumor, etc.) across the web is driven by user interest, modulated by daily and weekly cycles, and dampened over time. To adequately capture this behavior, this paper modifies the classical epidemic model (*Eq. (1)*) by adding a sinusoidal function (*Eq. (7)*) and a power law (*Eq. (5)*); this modification requires seven parameters. Note that this is a fluid approximation, since $B(t)$ is an integer.

Unlike the classical model, there is no way of integrating the resulting differential equation, so the equation is solved numerically by discretizing time. The parameters can then be calibrated by fitting measured data points with the Levenberg-Marquardt algorithm (a form of gradient descent to minimize least square errors).

Since measurements are needed for calibration, the model has limited predictive power (see *Sec. 4.4* for the tail and *Sec. 5.1* for the rise). Nonetheless, the parametric values serve to succinctly characterize the diffusion, and provide some insight (*Fig. 9*).

C H A P T E R   9

# Experimental Validation and Analysis

Every analytical model of a computer system needs experimental measurements to validate it. These experiments are often used to analyze the system as well. This chapter discusses some issues in model validation and experimental analysis.

To make the discussion coherent, we relate the issues to a case study, namely the impact of database locks on transaction performance. We first describe this problem.

## 9.1   CASE STUDY: DATABASE TRANSACTION LOCKING

A *database* is a collection of interrelated data. A *database management system* is a database together with a suite of programs for organizing, updating and querying the database. Users access a database through application programs that run on top of the database management system. Examples of such applications are airticket reservation and Internet auctions.

Database management systems have three important features: *persistence*—if a program modifies some data, the changes remain after the program has terminated; *sharing*—more than one program can concurrently access the data; and *reliability*—the data must remain correct despite hardware and software failures.

In manipulating data, a program may cause the database to be temporarily incorrect. For instance, in transferring an amount of money from an account $A$ to another account $B$, the database would be incorrect in the interval after the amount is deducted from $A$ and before it is added to $B$ (since that amount is missing from the total of the two accounts for the duration). The persistence of changes and this possibility of a temporary inconsistency lead to the requirement that either all changes made by the program up to its *commitment* (i.e., successful termination) is reflected in the database, or none at all. This is the concept of a *transaction*.

Since a database system allows multiple transactions to be active at the same time, the actions of transactions on shared data are interleaved. To prevent this interleaving from producing inconsistent data, there must be a *concurrency control* protocol to coordinate the transactions' actions.

The classical technique for concurrency control is *locking*: before a transaction can read or write on a piece of data, the transaction must set a readlock or a writelock (accordingly) on it. If another transaction already holds a lock on that data, and one of the two locks is a writelock, then there is a *conflict*, which must be resolved in some way. Conflict resolution is central to the problem of modeling how transaction locks affect performance.

There are two ways of resolving a conflict: (1) one of the two conflicting transactions is *aborted*—all its changes to the database are undone, its locks are released, and it has to start all over; or (2) the transaction that is trying to set the lock is *blocked*—it joins a queue of transactions waiting for the lock to be released. This blocking may lead to a *deadlock*, where several transactions wait for one another, unable to proceed. Such a situation must be detected and removed by aborting one of the transactions.

Concurrency control is just one factor among a myriad others that influence system performance, but the complexity in modeling this one factor is already evident from the brief description above. The dependencies involved and the sheer size of the problem (some applications have massive amounts of data and thousands of concurrent transactions) are mind-boggling. An exact stochastic analysis would be completely intractable—one could not begin to write down the state space.

Performance-related questions arise naturally once we look into the details of locking. For example, the division of secondary memory into pages makes it most convenient to lock a page at a time, even if a transaction's reads and writes refer to records, and a page contains more than one record. This brings up the question of *granularity*: How much data should be locked at a time? Intuitively, locking more would reduce the level of concurrency. On the other hand, it would reduce the number of locks a transaction needs, and thereby reduce the locking overhead. How does granularity affect performance?

Moving to a more abstract standpoint, which is a better way of resolving a conflict: blocking the transaction requesting the lock, or aborting one of the conflicting transactions? In aborting a transaction, all the work that is already done by the transaction would be wasted, whereas in blocking a transaction, the work that is done by that transaction is conserved. It would thus seem that we should always block the requesting transaction, rather than abort one of the two transactions. Is that correct?

When there is a deadlock, which transaction should we abort to break the cycle? One could, say, abort the transaction in the cycle that has consumed the least amount of resources. A transaction in a deadlock may be blocking a few transactions, so one could also argue for aborting the transaction that is blocking the largest number of transactions. There are several other possible criteria for choosing the victim. Which is optimum?

It is possible to avoid deadlocks altogether. Suppose a transaction is not allowed to begin execution if any of the locks it needs is already granted to some other transaction. It must wait till all those locks are available, get them, then begin execution; thereafter, it is not allowed to ask for more locks. This is called *static locking*, in contrast to *dynamic locking*, where transactions get their locks as they are needed. Intuitively, static locking has the disadvantage that transactions tend to hold on to locks for a longer period than if they acquire the locks only when necessary. The advantage is that once a transaction gets its locks, it will not be blocked (so there will be no deadlocks) because it will not ask for more locks. Which is the better policy?

Fundamental to the problem of locking performance is the fact that locking is necessary only because transactions are run in a multiprogramming fashion. With multiprogramming, there are two kinds of contention. There is the usual *resource contention*—e.g., queueing for processor cycles and I/O and, in the case of multiprocessor systems, contention over memories and buses. Added to this is *data contention*—conflicts over data that result in lock queues and transaction abortion. Each form of contention degrades system performance. What is the effect of each, and how do they interact?

## 9.2 MODEL VALIDATION AND EXPERIMENTAL ANALYSIS

We now discuss some issues in model validation and experimental analysis, using the case study above for illustration.

### 9.2.1 THE NEED FOR VALIDATION

To formulate a model for a complicated system with a limited number of equations and variables, we must inevitably make assumptions and use approximations. However, we must make sure that these assumptions and approximations do not remove essential features of system performance (e.g., linearizing what is actually nonlinear, or using a monotonic function to model some nonmonotonic behavior). Hence the need to validate the model with experimental measurements.

There is another, related reason for experimental measurements: There are often alternative ways of modeling the same system, using different techniques. For example, there are models for transaction locking that use Markov chains, and others that use queueing networks. In favoring a particular technique and model, we therefore need some supporting evidence from experiments.

Even for the same model and technique, there is usually a choice of whether to adopt a particular assumption, or how to make a necessary approximation. In fact, experimental measurements can help us make the choice.

In the case of database locking, for example, should a transaction's own locks be taken into account in calculating the probability of a conflict? Not having to do so may greatly simplify the analysis. One could make the choice by developing the model both ways, and using simulation measurements to see if the simplification causes a significant loss in accuracy.

## 9.2.2    DATA PRESENTATION

In presenting experimental data to validate a model, one should bear in mind the approximations in the model and the accuracy of the experiment. For example, it is not meaningful to use multiple decimal places when such precision is unwarranted.

It is often easier to compare numerical values with a data plot. However, one's judgement can be skewed if the axes are translated or scaled. For instance, a vertical axis that does not start at 0 can exaggerate the performance difference between two algorithms, and a logarithmic axis can understate the difference between model prediction and experimental measurement.

On a related note, the difference between model and experiment can also be reduced by choosing a closed, instead of an open, model. This is an issue that is examined in *OpenClosed* [46].

## 9.2.3    REAL SYSTEMS AND WORKLOADS

Ideally, a model should be validated with the real system itself—for example, the experiments in *InternetServices* [63] used actual implementations of the servers and applications. Unfortunately, this is not always possible. Database administrators loathe experiments with production systems, and vendors zealously guard against publication of performance measurements of their products.

One possible compromise is to log the events in a real system, then replay them on a simulator. Replaying such a trace would be more realistic than running a synthetic workload (like the TPC-C benchmark used in *SleepingDisks* [68]). One could even massage the trace by running it at a different speed, or cutting it and playing different pieces concurrently, etc.

However, the realism is also a weakness. For example, the trace from a database system is already filtered through its concurrency control, possibly altering the spatial and temporal correlation in the trace. One may therefore get misleading results from playing it on another system, with a different concurrency control, or at a different multiprogramming level.

A workload usually includes a dataset; sometimes, this dataset is necessarily synthetic. For example, if the experiments seek to test a system with a projected, larger dataset, then part or all of that dataset must be artificially generated. Even so, the synthetic data needs to be similar to real data. The problem of synthetically generating realistic data is difficult, and remains open [53].

## 9.2.4   SIMULATION

Given the difficulty of experimenting with a real system and the issues with using traces, it is not surprising that most models are validated by simulation. Since the simulator is playing the role of the real system in checking the model, the simulator needs to be as realistic as possible. A simulator that adopts the same assumptions (e.g., exponential distribution) and approximations (e.g., negligible overhead) does not serve its purpose.

Some simulation parameters may be calibrated with *microbenchmarks*. For example, disk service time may be measured by issuing a long sequence of IO requests. Such microbenchmarks are sometimes used also to validate a model but, even if they are run on a real system, such intense repeated activity may lack the realism required for validation.

## 9.2.5   PARAMETER SPACE REDUCTION

The more realistic a simulator is, the more parameters it has. Some parameters may be bounded; for example, $0 \leq u \leq 1$ for some utilization $u$. Other parameters may not be bounded, thus making an exhaustive exploration of the simulator's parameter space infeasible.

To reduce the parameter space, some subset of parameters are often fixed throughout the experiments with "magic values," so they become "voodoo constants." For example, there is one processor and two disks, or think time is $Z = 1$ s. Such constants may affect the generality of the conclusions drawn from the experimental analysis.

Sometimes, parameters are instantiated through *normalization*; e.g., *BitTorrent* [44] assumes "without loss of generality" that file size is 1. Such normalization can obscure the impact that a parameter has on performance. For example, setting $L = 1$ can prevent one from seeing an $L^2$ factor in a performance measure.

One analytically sound way of reducing the parameter space is to identify a subset of parameters $S$, and show that performance depends on $S$ only through some aggregate parameter $W$ that is expressible in terms of $S$. In the case of transaction locking, for example, if multiprogramming level is $N$ and database size is $D$, then one can show that (under certain assumptions) the conflict probability, response time and restart rate depend on $S = \{N, D\}$ only through $W = N/D$. In effect, this reduces the parameter space by one dimension.

## 9.2.6   UNINTERESTING REGIONS OF PARAMETER SPACE

As one increases the workload, system performance must eventually saturate or degrade. There is then no point in experimentally exploring the parameter space beyond that. For example, if transaction length is $k$, then again one can show that there is a constant $c$ such that throughput degrades when $k^2 N/D > c$, where $c$ depends on the assumptions. Attention can hence focus on the region where $k^2 N/D \leq c$.

However, one should be careful about using saturation to delimit the parameter space. For example, there is a huge literature on modifying TCP's congestion control, and throughput degrades when there is congestion, but one would miss the point of those modifications [59] if the parameter space were limited to where there is no throughput degradation.

The parameter space can also be delimited by performance measures themselves. For example, one can reasonably say that, for transaction locking, a workload is of no interest if the resulting conflict probability exceeds 0.5. Similarly, one can use realistic wide-area packet loss rates to rule out large areas of parameter space for Internet traffic.

There is another reason for such performance-based confinement of the parameter space: any experimental analysis from beyond would be of no practical interest. For example, to study which transaction in a deadlock cycle should be chosen for abortion, one would need to have workloads that generate large cycles. However, it is known that such workloads would drive the system far into its performance degradation zone; this implies that victim selection for large cycles is, in fact, a non-issue for database transactions.

At the other extreme, some regions of parameter space are uninteresting because the workload is light and performance is linear. Performance modeling for light workloads should be a straightforward exercise. The challenge lies in crafting a model that captures nonlinear behavior.

## 9.2.7   QUANTITATIVE PREDICTION VS. QUALITATIVE UNDERSTANDING

One reason for having an analytical model is to use it for performance prediction (e.g., *SleepingDisks* [68] uses a queueing model to dynamically decide disk speed), but there is a common misunderstanding that prediction is the only reason. For example, a model for transactions with a nonuniform access pattern over a database of size $D$ may yield an equation that says the performance is equal to having uniform access over a database of size $\alpha D$, where $\alpha$ depends on the access pattern and is not known a priori. Such an equation can be validated through experiments, but what is the point of doing so, if such an equation cannot be used to predict system performance?

However, an analytical model is not rendered useless if its results cannot be used for quantitative prediction, as they may help us understand qualitatively the behavior of the system. To see the difference, consider: No parent expects to predict a child's behavior, but all parents seek to understand their behavior.

If a model can help us understand how a system should behave, that may suffice for, say, configuration debugging and anomaly detection.

## 9.2.8    ANALYTIC VALIDATION

Analytic modeling is an art, in that we pick approximations to trade accuracy for tractability and insight. A different choice of approximations would give different expressions for the performance measures. Hence, theoretical conclusions from an analytic model must be validated to make sure that they describe properties of the system, rather than properties of the model. I call this **analytic validation**.

Instead of comparing experimental measurement and model calculation, one can do an analytic validation by comparing experimental measurements. For example, if the model shows that the performance of transactions that set read and write locks for a database of size $D$ is equivalent to the performance of transactions that set only write locks for a database of size $\beta D$ (for some $\beta$ that depends on fraction of write locks, access pattern, etc.), this is a result that calls for analytic validation. One can do this by running two experiments—one with read and write locks for a database of size $D$, and one with write locks for a database of size $\beta D$; an agreement in transaction performance from the two experiments then serves to analytically validate the model.

Is analytic validation necessary? Does it not suffice to have a numerical validation to check that the model is accurate when compared to experimental measurements? Yes, if the model and experiments agree exactly, then there is no need for analytic validation. However, the approximations needed for tractability always imply some loss of accuracy (and one should be suspicious of any model that can provide exact agreement).

There is a subtle purpose to analytic validation: A reader is asked to suspend disbelief as the model adopts one approximation after another to get to the theoretical conclusions, and the analytic validation serves to return the reader's faith.

## 9.3   DISCUSSION OF PAPERS

We now examine the experiments in the papers that we discussed so far, and in the four papers introduced here (*WirelessCapacity* [21], *NonstationaryMix* [50], *P2PVoD* [16]) and *MapReduce* [65].

*StreamJoins* [28]

Parameters in the model are calibrated with microbenchmarks. For example, the cost $P_d$ for accessing one tuple in data structure $d$ during search is calibrated by measuring the total running time for 6000 tuples at 100 tuples/s. It is not clear if such a calibration is sensitive to the workload mix as when there are, say, multiple streams and processor cache contention.

The usual experimental validation of a model would show a table or plot where measurement is compared to calculation. Here, however, the measurement and calculated values are shown in two separate plots—*Figs. 2* and *3*. This is a form of analytic validation: The arrangement of lines obtained with the model is compared to the arrangement from experimental measurement.

The two plots are impressively similar, except for some nonmonotonic behavior in the measurement for small and large $\lambda_a/\lambda_b$ in *Fig. 3*. This behavior is possibly a hint that there is some nonlinear effects not modeled by the equations (i.e., some limit to their validity); if so, the subsequent plots *Figs. 4–7* may not show straight lines if the parametric values were different.

In *Eq. (8)*, the parameters $P_n$, $P_h$, $I_n$, $I_h$ are instantiated, and the ratio $B/|B|$ set to 10. This obscures some possibly interesting issues: What if $I_h < I_n$ or $P_n|B| < P_h$, etc.?

*SleepingDisks* [68]

There is limited validation of the model (*Fig. 3* and *Table 3*) using a one-day trace collected from an IBM DB2 database server running TPC-C like transactions (*Fig. 6*). Another trace used (Cello99) was from a UNIX file system.

For a stable workload, 4-h segments were cut from the traces and run at different speeds. It is not clear if any realism is lost through such a change in speed.

For a dynamic workload, the database benchmark was run with 10, 55, and 100 clients, and the traces replayed sequentially. This preserved any correlation among the disk accesses that might have been imposed by DB2. In contrast, concurrently running 10 copies of the 10-client trace, say, would probably not produce access behavior similar to that for the 100-client trace.

In using the same disk trace to compare energy saving schemes, there is an implicit assumption that the schemes would not affect the access pattern. This may not be so. For example, with the database workload, an increase in disk response time under one scheme may cause a lock to be held longer by the transaction from one client, possibly causing a delay in disk access for a conflicting transaction from another client.

*GPRS* [42]

Although the experiments simulated the radio resource allocator, they adopted the same ON/OFF traffic and memoryless assumptions as the model. The close agreement between model and measurement in *Figs. 5* and *6* is therefore not surprising.

*Table 1* sets $T = 4$ and $d = 4$, so $\min(nd, T)$ is either 0 ($n = 0$) or 4 ($n \neq 0$), possibly restricting severely the range of experimental results.

The plots show that some experiments pushed the blocking probability to levels that are unrealistic, considering that the maximum acceptable blocking probability is assumed to be 2% (*Sec. 6.1*); one must therefore take care in drawing conclusions from those experiments.

*TCP* [43]

The TCP that one finds in current operating systems is an intricate protocol, and the Internet it runs over is even more complicated. Even so, the authors have found a way to capture the essence of TCP behavior, and provide convincing validation (*Sec. 3*) that *Eq. (31)* from the simplistic model matches measurements from the complex system.

*BitTorrent* [44]

The validation in *Secs. 6.1* and *6.2* is a comparison of the deterministic fluid approximation to a discrete-event simulation of the stochastic fluid differential equation (*Sec. 3.4*). Although the two models are similar in assumptions, the plots show that the simulation measurements can vary considerably from the deterministic curve, although the equilibrium is stable.

Validation with the real system is more difficult. For the fluid approximation to work well, the seed must attract a large number of downloads. The comparison between fluid model and BitTorrent measurement in *Sec. 6.3* is therefore inconclusive, since the seed (for copyright reasons) is not popular.

*CodeRed* [69]

*Fig. 9* shows good agreement between the observed number of infected hosts and the two-factor model. Some tuning of the parameters ($\gamma$, $\mu$, etc.), is necessary, but it confirms that *Eq. (17)* does fit the data.

This is an example where the model cannot make quantitative predictions (since the parametric values depend on the particular malware), but it is useful for understanding the system.

*MediaStreaming* [62]

For the model validation in *Table 1*, the largest error is about 15%. Such accuracy does not warrant using four significant figures in the numerical values.

One could do a validation of the crucial exponential growth equation, as follows: *Eq. (2)* has equivalent form $\log(N + \alpha P(k)) = k \log(1 + \frac{\alpha}{b}) + \log N$; measure $P(k)$ and plot $\log(N + \alpha P(k))$ against $k$ for fixed $\alpha$, $b$ and $N$; linear regression on these data points should give a line with gradient $\log(1 + \frac{\alpha}{b})$ and vertical intercept $\log N$.

Although the model fixes $L$ and $b$, and discretizes time, the validation would be stronger if the experiments randomize $L$ and $b$, and do not divide time with $L$. One can then validate *Eq. (3)* as follows; use the equivalent form $k_0 \log(1 + \frac{\alpha}{b}) = \log(\frac{\lambda Lb}{N})$; measure $k_0$, average $L$ and average $b$ for various combinations of $\alpha$, $\lambda$ and $N$; for each combination, plot $k_0 \log(1 + \frac{\alpha}{b})$ against $\log(\frac{\lambda Lb}{N})$. These data points should scatter along the 45° diagonal. This would be an example of analytic validation, where a theoretically derived relationship (*Eq. (3)*) is validated by using experimental measurements that are independent of the model.

Rather than fix $L$ and $b$ to reduce the parameter space, one could define another parameter $C = \frac{\alpha}{b}$ and system workload $W = \frac{\lambda Lb}{N}$. The key result in *Eq. (3)* depends only on $C$ and $W$, and this 2-dimensional parameter space can be explored exhaustively.

*StorageAvailability* [18]

The paper validates the model with a simulation. *Fig. 3* shows impressive agreement between simulation measurement and model calculation, but there are not much details about the simulator. While it may implement various details of StarFish (write quorum, HE failover, etc.), it may also use memoryless distributions to generate failure and recovery.

In using microbenchmarks to measure performance, one must be mindful that they don't behave like a real workload. For example, random reads would not have the same latency as reads generated by a sequential scan or other forms of spatial and temporal locality. In their experiments, the authors take care to arrange for the reads to be all cache hits or all cache misses, and thus calibrate the hit and miss latency with *Table 6*.

Since there is a tradeoff between availability and latency, the availability results in *Table 1* and the latency results in *Table 5* should have been integrated and plotted (e.g., latency vs. availability, one data point per configuration) so one can explicitly see the tradeoff.

*TransactionalMemory* [24]

The authors use the aggregated parameter $kN/L$ to delimit their parameter space. As previously noted (Chapter 3), the model is similar to the no-waiting case for database locking [58], for which it is known that performance is determined by $k^2 N/D$. This result appears to hold here as well: the authors note that doubling $k$ quadruples $E(R)$ (*Table III*), and decreasing $L$ by a factor of 10 increases $E(R)$ by a factor of 10. Hence, the authors should perhaps have used $k^2 N/L$ to delimit their parameter space; much of the discrepancies in *Figs. 2–4* may lie outside that limit. For the no-waiting case, the region of interest is $k^2 N/D < 6$, identified by considering throughput, which decreases when $k^2 N/D > 6$.

*SensorNet* [49]

Consider *Eq. (5)*, which says expected MULE buffer occupancy is $E(M) = \rho_{sensors} N^2$. Why should buffer occupancy depend on $N$, which is an arbitrary parameter? Surely, if we refine the grid (e.g., use $4N$ instead of $N$), that should not affect $E(M)$? In contrast, the expression $\rho_{sensors}/(\rho_{AP}\rho_{mules})$ in *Eq. (18)* is independent of $N$.

Here, the time and space discretization are too tightly related. Suppose the model includes another parameter $s$, and at each clock tick, a MULE takes one step in the grid and and each sensor generates $s$ units of data. The expressions for performance measures would then include $s$, and would scale appropriately if the grid size is changed. By fixing $s = 1$, the model adopts a normalization that introduces an artifact (performance depends on grid size).

The experiments use the same space/time discretization as the model. The validation would be stronger if the experiments could show that the results still hold even if the modeling assumptions were relaxed. For example, the model assumes that APs are spaced exactly $\sqrt{K}$ grid points apart; this assumption is used for the folding argument in *Fig. 4*. However, the experiments show that the numerical results are similar even if the APs are randomly placed—this is a validation of the model that deserves to be more than just a footnote (*Sec. IX*).

Similarly, while a discrete-time random walk simplifies the analysis, the experiments could have used an alternative mobility model [4] over a continuous space. Although many mobility models are not realistic, if simulation with these models agree numerically with the analytical model, that would strengthen its validation.

The authors describes the curves in *Figs. 10* and *11* as "steep," forgetting that the horizontal axes are logarithmic.

*NetworkProcessor* [34]

The SpliceNP throughput calculation with the queueing network model in *Fig. 4* shows impressive accuracy, especially since the comparison is with a real implementation.

*Fig. 5* is a typical example of the $T$-vs-$\lambda$ curve illustrated in Fig. 2.5 for an open system.

*DatabaseSerializability* [3]

The interaction between resource contention and data contention can be seen in this paper: An increase in data contention (e.g., raising the multiprogramming level) causes more blocking, so that may reduce resource contention; on the other hand, the increased blocking causes more freshness violation, and more aborted transactions and wasted resources.

Conversely, an increase in the number of caches makes it more likely that a transaction can access a local copy, possibly reducing data contention; on the other hand, more caches also implies more copiers, and that may lead to higher data contention. *Fig. 12* shows that, with BSAFC-0, this data contention when scaling out (i.e., increasing the number of caches) can cause the throughput to drop, possibly to below that for no caches.

## *NoC* [29]

*Fig. 5* shows good agreement between model and simulation, even as latency increases sharply. This is despite the many approximations (memoryless service time, use of $M/M/1$ formulas for $M/M/1/D$, etc.). It also means that the power consumption linearity and agreement in *Fig. 6* are not because $P_{pe}$ is small. However, the linearity also suggests that the nonlinear terms, like $BP_{leak\_flit}$ in *Eq. (26)* are made negligible by the small value of $P_{leak\_flit}$. That means the elaborate queueing model is unnecessary for estimating power consumption.

In analytical modeling, it is often useful to first run some experiments (with a simulator, say) to determine what aspects of the system the model should focus on, before formulating the model.

Note that the power consumption parameters ($P_{r\_buf}$, etc.), in the simulator are calibrated with a hardware implementation.

## *WebCrawler* [8]

*Fig. 9* shows a key result from the model that there is some threshold $\lambda_h$ on the update rate $\lambda$, beyond which the crawl frequency $f$ should decrease as $\lambda$ increases. In reality, the value of $\lambda_h$ may be unknown and impossible to predict, but the point of the result lies in the qualitative insight that the optimal $f$ is not a monotonic function of $\lambda$.

*Table 2* is generated by a synthetic workload (all pages change with an average 4-month interval). It would be more convincing if a trace is used instead.

## *DatacenterAMP* [22]

In the experiments, the service time is controlled by varying the size of matrices and the number of cores. However, the inter-arrival time is exponentially distributed, like in the model. The validation would have been stronger if the experiments use an arrival process that is different from the model, e.g., by using some trace data that includes bursty arrivals. That would show that the model's conclusions are more general than its Markovian assumption.

## *RouterBuffer* [1]

The authors' expression for $E[Q]$ and $\text{Prob}(Q \geq B)$ depend on the link rate $C$ only through utilization $\rho$. These are strong claims from the model, so they require experimental verification. Indeed, *Fig. 9* and *Fig. 11* show that, when $\rho$ is fixed, the $E[Q]$ and buffer requirement for different $C$ are indistinguishable. These are examples of analytic validation.

In TCP's additive-increase and multiplicative-increase phase, a dropped packet prompts the sender to reduce $W_{max}$ by a multiplicative constant $\alpha$. The authors follow current practice by fixing $\alpha = 1/2$ in this paper. This hides a nonlinear effect that $\alpha$ has on buffer sizing.

*Sec. 3.2* uses the model to compute utilization bounds for various buffer sizes, but it is not clear if the many decimal places are meaningful (e.g., Util$\geq$ 99.99997%). For example, a comparison of the lower bounds in the **Model** column of *Fig. 14* and the values in **Sim.** and **Exp.** columns show that the bounds are violated by the measurements.

## *PipelineParallelism* [41]

*Fig. 6* compares the measured execution time for the Pthreads implementation to that calculated with the queueing models. The agreement in *Fig. 6(a)* is excellent.

However, *Fig. 6(b)* shows that the execution time does not decrease monotonically as the number of threads increases, but a separable queueing network does not have such non-monotonicity. This is an example where some property of the solution (here, monotonicity) is in fact an artifact of the model.

We mentioned in Section 9.2.1 that experiments are needed not just to validate a model, but to guide its formulation. Here, if we first run the experiment in *Fig. 6(b)* and observe that nonmonotonicity, that should give us pause before adopting a queueing network model.

Recall our observation in Section 9.2.2 that the difference between model and experiment is generally smaller if the system is closed, instead of open. We see this effect in *Fig. 6*, where the closed `ferret` model is more accurate than the open `dedup` model.

*Sec. 3.2* concludes that three stages are optimal, i.e., all the intermediate pipeline stages should be collapsed into one, in which case there is no need for an analytic model to decide how to structure the pipeline and threads. This conclusion suggests that the analysis may be looking at some uninteresting region of the parameter space.

*Sec. 3.4* uses the models for numerical calculation and thus determine the optimum configuration of the pipeline (number of stages and threads). It is also used to numerically verify that work stealing is close to optimum (*Fig. 9*, which is also a validation of the use of virtual servers to model work stealing). The analytical model thus verifies that collapsing stages and work stealing can overcome work imbalance in the pipeline.

## *Roofline* [66]

This is an example of an analytic model that is used for quantitative prediction. The roofline bounds are validated by executing the kernels on the hardware. The 16 combinations in *Table 4* are marked as crosses in *Figs. 3* and *4*. The bounds for `FFT` are particularly loose.

*Gossip* [2]

Fig. 5 uses a single plot (one each for $n = 500$ and $n = 2000$) to compare measurements of $x(t)$ from a simulation of the protocol to the value calculated with *Eq. (4)*. This is a straightforward numerical validation of the model.

In contrast, *Fig. 4* presents protocol measurements and model calculation in separate plots. The striking similarity in the patterns is an analytic validation—it shows that the model does capture the *behavior* of the protocol (even if there is significant numerical discrepancy between measurement and calculation).

*EpidemicRouting* [67]

We see an excellent example of parameter space reduction in *Eq. (1)*, where the mobility model is reduced to an exponential distribution with the single parameter $\beta$. This simplifies tremendously the closed-form expressions in *Table 1*.

The validation in *Figs. 1* and *2* shows good agreement between simulation and model (except *Fig. 2(c)*), but the results would be more persuasive if the simulator does not copy the model's assumptions. For example, instead of assuming packet generation by a Poisson process, the simulation could have included traffic that has constant bit rate.

*SoftErrors* [40]

*Sec. 5.3* notes that an aggregate metric like IPC does not correlate with AVF, but the supporting data (*Fig. 8a*) uses CPI (cycles per instruction). IPC and CPI are reciprocals of each other, but their relative errors are different. For example, if the CPI relative error is $\frac{C'-C}{C} = \frac{C'}{C} - 1 = 0.5$, the corresponding IPC relative error is $(\frac{1}{C'} - \frac{1}{C})/(\frac{1}{C}) = \frac{C}{C'} - 1 = \frac{1}{1.5} - 1 \approx -0.3$.

*Eq. (3)* shows that occupancy calculation does not depend on average instruction latency $\ell$ and dispatch width $D$ separately. Rather, it can be expressed in terms of an aggregated parameter $A = \ell D$, thus reducing the parameter space by one dimension.

*GPU* [26]

The authors fix the issue rate at 1 instruction per cycle, thus removing one dimension from the parameter space. However, they indicate in the equations where this assumption can be relaxed. To illustrate the scalability of `cfd_step_factor`, the paper should present a plot of IPC, in addition to the CPI in *Fig. 16*.

*ServerEnergy* [19]

ACES uses some magic values—$\alpha = 0.95$, $s_{\max} = 100$, etc.

*DatabaseScalability* [14]

Although much effort was put into estimating abort probability $A_N$ for multi-master replication, the benchmarks in the experiments have low abort rates, and so do not stress the estimates. The low abort rates indicate that the system's performance is largely determined by resource contention, not data contention.

The authors stressed their model by artificially increasing the abort rate (*Sec. 6.3.3*). As they pointed out, an abort rate of 29% would be intolerable. The inaccuracy in predicting $A_N$ in *Fig. 14* for a large number of replicas is thus unimportant. However, even a small divergence between prediction and measurement for acceptable values of $A_N$ can translate into a large discrepancy between prediction and measurement in other metrics, especially response time (which, unlike throughput, is unbounded).

Standard MVA assumes service demand does not increase with job population. The throughput therefore increases monotonically toward the bottleneck bound. This paper, however, scales the demand by a factor $1/(1 - A_N)$ that grows with the number of clients, so the predicted throughput would eventually drop as the number of replicas increases. The measured throughput should behave similarly as abort probability increases. If the model prediction can match this nonmonotonic behavior in measurement, that would represent a triumph of the model.

The squiggly lines in the plots are odd. They appear to be drawn by some software that tries to connect given data points, and are misleading. One should simply plot just the data points themselves, so the reader can see the irregularity as just statistical fluctuations.

*PerformanceAssurance* [45]

The horizontal axis in *Fig. 8* has an unusual scale (not linear, not logarithmic) that makes it difficult to parse the plot. The multiple decimal places (e.g., 45.1062) in *Eq. (2)* and *Eq. (3)* are not meaningful.

*Fig. 11* illustrates the point that it is easier for a model to be accurate for throughput and utilization (which are bounded) than for response time (which is unbounded).

*CachingSystems* [36]

*Figs. 2–4* show excellent agreement between model and simulation, but the simulation adopts the same assumptions as the model (IRM for *Fig. 2* and hyperexponential for *Fig. 3* and *Fig. 4*). It would be more interesting to see if the model can match the hit probability for the Youtube traces in *Fig. 5*.

*Sec. IV-G* presents small-cache approximations for the hit probability. One should verify that the policies have the behavior specified by these closed-form expressions derived from the model, i.e., an analytic validation. In the case of $p_{\text{hit}}(m) \approx (\lambda_m T_C)^k$ for $k$-LRU, say, one could take measurements of $\langle x, y \rangle = \langle \lambda_m T_C, p_{\text{hit}}(m) \rangle$ and verify that the data points $\langle \log x, \log y \rangle$ lie around a straight line through $\langle 0, 0 \rangle$, with gradient $k$.

*InformationDiffusion* [37]

It is unclear why the exponent in the power law (*Eq. (5)*) and the period $P_p$ are not included in the parameter set $\theta$. It is also not obvious why the model is depicted as a thick line in *Fig. 8* (especially considering the log scale), nor why the model is needed to identify the outliers there.

### *WirelessCapacity* [21]

The system consists of stationary wireless nodes that are geographically distributed. The spread is bigger than the transmission range, so traffic between nodes may require multiple hops. The channel is shared, and senders contend for medium access.

The issue is scalability: How does the number of nodes affect the achievable throughput per sender? The main parameters are the number of nodes $n$ and channel bandwidth $W$; other parameters include transmission range $r$, power $P_i$ for node $i$, and signal-to-interference ratio $\beta$. Performance is measured with transport capacity (bit-meters) for all $n$ nodes, and per-node throughput $\lambda(n)$.

We see here an example where, for the same performance problem, the results that one gets depend on the model for the system. For example, with an Arbitrary Network Model, the Protocol and Physical Models give different upper bounds on the transport capacity (*Theorem 2.1*). Unfortunately, the authors offer no experimental validation for the results. This is understandable, since a simulator would make similar assumptions about interference, while a real wireless network would be hard to configure and scale up.

Real wireless communication is notoriously flaky. Therefore, in this case, there is perhaps no point in sweating over the detailed differences in the results between Arbitrary and Random Networks and between Protocol and Physical Models. After all, the implications and tradeoffs that one deduces from these models (*Sec. I.C and I.D*) are not sensitive to these differences.

Even so, one should be careful how some detail in the model can affect the result. For example, how should the model specify the relationship between source and destination? Suppose the destination is "randomly chosen," using the Random Network Model. For a fixed transmission range, the impact of this assumption on scalability depends on how topology changes with the number of nodes $n$: If density is constant, then area $A$ increases with $n$, thus increasing the number of hops between source and destination; if $A$ is fixed, then the density increases with $n$, thus increasing the number of nodes per hop that suffer interference. Therefore, one should avoid the normalization $A = 1$, so that its appearance in the formulas helps to make the difference in impact explicit.

### *NonstationaryMix* [50]

A challenge that one faces when modeling a real system is that parameters that one routinely use to construct a model may not be available. For example, service times may be unknown because they require possibly intrusive instrumentation of a production system.

One solution would be to device some indirect calibration of the parameters. However, real systems usually run a mix of transaction types, and the mix can vary drastically over time. How can the parameters be calibrated when the mix is nonstationary?

These issues are common in other computing systems. When modeling Internet traffic, for example, one often have application data at the end hosts only (no router measurements, etc.), and the traffic mix is constantly changing.

The authors' idea for addressing this issue lies in the insight that the nonstationarity also provides a collection of independent data points (e.g., a sampling from the set of points in *Fig. 1*) that can be used to calibrate model parameters by linear regression. This provides an alternative to calibration by controlled experiments (with benchmarks, say).

The paper considers an enterprise system running a mix of transactions. The model discretizes time into 5-minute intervals. The model parameters are $N_{ij}$ (the number of type $j$ transactions in the $i$-th interval), and the regression coefficients $\alpha_j$ and $\beta_{jr}$. These coefficients are obtained by regression using $N_{ij}$ and response time $y_i$ (*Eq. (2)*) and utilization $U_{ir}$ (*Eq. (5)*); these values are usually available from routine system logs.

The Basic model starts with $T_{ij} = \alpha_j N_{ij}$. Recall from Sec. 6.1 that $T \approx \sum_r D_r$ when there is little queueing (cf. *Assumption 4* in *Sec. 3.1*). Since $\alpha_j$ is sum of service times over all resources $r$ and $D_r$ is product of number of visits and service time per visit, the Basic model works best if each resource is visited just once, like in *Fig. 6*. One interpretation of this is that, for each resource $r$, the model adds the service times for multiple visits to that resource, and consider the sum to be the service time for just one visit. This rearrangement of visits will give similar performance if there is little queueing on any visit.

Although the emphasis of this paper is on nonstationarity, the Extended model (*Eq. (4)*) uses the steady-state $M/M/1$ waiting time $\frac{1}{\lambda_i} \frac{U_{ir}^2}{1-U_{ir}}$ (see Eq. 2.7) to estimate $y_i$. This recalls the idea of flow equivalence (see Sec. 6.2): although the transaction mix arriving at the system (represented by the $M/M/1$ queue) is nonstationary, the approximation may work well if the system reaches equilibrium within each 5-minute interval.

The Composite model seeks to replace the performance metric $U_{ir}$ in the Extended model with an expression in terms of input parameters. If $\lambda_{ij}$, $V_{ijr}$ and $S_{jr}$ are the arrival rate, visits per job and service time per job ($i$, $j$, $r$ as before), then

$$U_{ir} = \sum_j \lambda_{ij} V_{ijr} S_{jr} = \sum_j \frac{N_{ij}}{L} V_{ijr} S_{jr} = \sum_j \frac{V_{ijr} S_{jr}}{L} N_{ij},$$

so the service demand $\beta_{jr}$ in *Eq. (5)* assumes $V_{ijr}$ does not change with interval $i$.

Even when two isolated workloads are well understood, predicting their performance when run together is difficult because performance measures like queue lengths and response times cannot be simply added together. However, arrival rates and therefore utilization can be added. Since the Extended model uses utilization, it easily leads to the Consolidated model in *Sec. 5.1*. Nonetheless, this model would break if $U'_{ir} + U''_{ir} \geq 1$; here, the model again relies on *Assumption 4* (i.e., over-provisioning).

While aggregate errors (e.g., *Table 2*) and cumulative distributions (e.g., *Fig. 11*) give an overview of the model's accuracy, further details are necessary for a better understanding. For example, a large error in predicting sub-second response times (e.g., PetStore in *Table 1*) may not matter, and a small error in predicting a high utilization of, say, 0.98 is easy to achieve (since it is a bottleneck).

*Fig. 8* illustrates how an analytical model can be used to identify bugs when measured performance deviates significantly from prediction.

### P2PVoD [16]

The runaway success of P2P systems like BitTorrent has prompted many proposals from academia to improve such protocols. Each proposal can be viewed as a point in the design space. This paper proposes an analytical model to represent the entire space defined by the three dimensions of throughput, sequentiality and robustness. The tradeoffs in choosing one point instead of another in this space can then be analyzed via the model.

The system consists of videos that are divided into chunks, a seed that has all the videos, and peers; of the network topology, only the access link is considered. These are specified with the number of chunks $M$, the chunk size $b$, the uplink capacity $U_s$ for the seed and $U_p$ for a peer, and the peer arrival rate $\lambda$.

For the metrics, robustness is defined as $R = 1 - p^{\bar{r}}$, where $p$ is the probability that a peer is bad and $\bar{r}$ is the average number of sources per chunk. Sequentiality $S$ is defined by *Eq. (10)*, and throughput measured with downloading time $T$. Buffering time is central to this paper, but the authors do not model it.

For a queueing network, the bottleneck queue can be used to derive an upper bound on throughput and a lower bound on delay (Fig. 6.2). Although the system here is not modeled as a queueing network, the idea remains applicable: the authors use the bottleneck chunk to derive bounds on $\lambda$ and $T$ (*Eq. (7)* and *Eq. (8)*). Incidentally, *Eq. (4)* uses another idea from queueing networks, namely operational analysis [10].

In the experiments, peers join the swarm at a uniform spacing of $\lambda$ peers per minute. It would be more realistic to have a randomized inter-arrival time. The model's inaccuracy in *Fig. 4* shows that one should be careful about accepting the closed-form expressions at face value. Part of the inaccuracy comes from the magic value of limiting the experiments to 50 nodes, which is not enough for steady state under heavy traffic.

*Sec. 7* mentions that this paper was developed via a back-and-forth between the analysis and the experiments. This bears out our point in Sec. 9.2.1 that the role of experiments is not just to validate a model, but also to guide its development.

*MapReduce* [65]

A MapReduce job consists of map, shuffle and reduce tasks that run on processors and disks in a distributed system. The job performance depends not only on the queueing delays for these resources, but also on the precedence constraints among the tasks. *Table 1* lists some input parameters for the architecture $(n, c, d)$ and the workload $(D_{i,k}, m, r, pm, pr, ps)$.

Modeling how precedence impacts task performance is a challenge. For example, such a constraint violates the assumptions for separable queueing networks. Nonetheless, this paper uses a hierarchical decomposition, where one level models the precedence constraint, and the other level is a multiclass queueing network model for the resource contention. In this queueing network, the $N$ jobs have identical precedence graphs, and each task in a job belongs to its own class (so each class has $N$ statistically identical tasks).

Without the precedence constraint, each task in a closed queueing network is regenerated when it completes, so any two tasks would overlap (in time) completely. With the precedence constraint, the overlap is reduced; this effect is modeled in *Eq. (6)* by having an arriving task see the time-averaged queue size $Q_{jk}(\overrightarrow{N - 1_i})$ reduced by factors $\alpha_{ij}$ and $\beta_{ij}$ for intra-job and inter-job task overlap. (The factors $\frac{1}{N}$ and $\frac{N-1}{N}$ model the probability that the arriving task overlaps with another task in the same job or in another job.)

The $\alpha_{ij}$ and $\beta_{ij}$ values are calculated by first constructing a precedence tree (*Fig. 6*), with the help of a time line (*Fig. 5*) that is laid out using (i) *average* response times calculated with the queueing model, and (ii) a simulation with *logical queues* (*Sec. 4.2.2*). Since solving the queueing model requires, in turn, the $\alpha_{ij}$ and $\beta_{ij}$ values, an iteration between the precedence and queueing models is used to get a fixed-point solution. This approximation uses *Eq. (5)*—an analog of Schweitzer's approximation—to break the MVA recursion.

The tasks are leaves in the precedence tree. After calculating the average response time of each task, the average response time of each subtree is calculated, bottom-up, to give the response time of the job at the root. For a $P_A$-rooted subtree, the paper considers two different methods from previous work for calculating the subtree's response time (*Sec. 4.3*).

The experiments show excellent agreement between simulated performance and the model's calculations, but the simulator assumes task response times are exponential and replaces each CPU and disk by a queue, like in the model. (On the other hand, the service demands in *Table 2* are from real measurements, but it is not clear how they are adjusted when the number of threads change in the experiments.) *Table 4* shows a wider gap between a real setup and the model. There is some spurious accuracy in the measurements (e.g., 37.4983 in *Table 2* and 722.23 in *Table 4*), and the use of *pm*, *pr* and *ps* is a poor choice of mathematical notation for model parameters.

The experiments run only 1 job at a time, so they do not test the model for inter-job synchronization ($N = 1$ in *Eq. (6)*). The demonstrated accuracy in the model is partly because the performance has hit a bottleneck: response time is mostly linear in *Fig. 7* and utilization is flat in *Fig. 8*. *Fig. 9* further shows that the system's configuration cannot optimally support more than a handful of nodes and tasks. (Incidentally, the vertical axis for *Fig. 9c* should start at 0, since it aims to show the accuracy of the model.)

This paper provides an example of how experiments are used to choose an approximation for the model; specifically, whether $\max\{EX_1, \ldots, EX_k\}$ can adequately approximate $E \max\{X_1, \ldots, X_k\}$ for estimating the response time of a $P_A$ subtree (*Sec. 4.3*).

CHAPTER 10

# Analysis with an Analytical Model

The previous chapter examines some issues concerning experiments. We now continue by discussing the formulation and function of analytical models.

## 10.1 THE SCIENCE AND ART IN PERFORMANCE MODELING

Where appropriate, we again use the case study on transaction locking as illustration.

### 10.1.1 POWER

Why have an analytical model? A common argument is that solving them takes less time than running simulations but (given the rapid increase in compute power) this claim is increasingly weak. Another argument is that some simulators are very complicated (e.g., the network simulator ns-2), and contain bugs that may affect the reliability of their results, but solvers for analytical models can have bugs too.

Although analytical models are often proposed as alternatives to simulation, many are in fact used as simulators, in the following sense: After validating the model with comparisons to experimental measurements, the model is used to plot the relationship between performance measures and input parameters, and conclusions then drawn from these plots. I call this **analytic simulation**.

If conclusions are to be drawn from looking at plots, then one may be better off generating those plots with a simulator, since the model's assumptions and approximations may introduce misleading inaccuracies.

The power in an analytical model lies not in its role as a fast substitute for a simulator, but in the analysis that one can bring to bear on its equations. Such an analysis can yield insights that cannot be obtained by eyeballing any number of plots (without knowing what you are looking for), and provide conclusions that no simulator can offer.

### 10.1.2 TECHNIQUE

The power of an analytical model is most obvious when it provides expressions that explicitly relate the performance measures to the system parameters. Such closed-form expressions can then be subjected to mathematical analysis. In the case of database locking, this was how the workload bound $k^2 N/D < c$ was discovered.

In contrast, although Markov chains are highly flexible, they usually do not provide closed-form expressions (e.g., *GPRS* [42]). Often, the performance metrics are only implicitly specified as the solution to a set of equations. This is also the case for queueing network models (e.g., *InternetServices* [63]). Such implicit solutions restrict the theoretical conclusions that one can extract for the system. Sometimes, some limited analysis may still be possible; we see this for the Kermack-McKendrick model (Section 8.2), and this is so for queueing networks as well [51].

Sometimes, a modeling technique may imply an artifact that is not a feature of the system. For example, the hardest part of modeling transaction locking is estimating the waiting time for a lock, and one possible work-around is to make a transaction repeat a conflicting request at most $L$ times with random intervals; if the requested lock is still unavailable after $L + 1$ tries, the transaction restarts. Unfortunately, this model [6] implies that $L$ should be 0, i.e., a transaction should not wait for a lock. Here, the memoryless request repetitions cannot model an essential aspect of the locking protocol.

### 10.1.3 ASSUMPTIONS AND APPROXIMATIONS

Every model starts with assumptions. Some of them serve to specify and simplify the system; e.g., two-speed disks (*SleepingDisks* [68]) and constant infection rate (*CodeRed* [69]). Some others are used to facilitate the derivation; e.g., packets lost in one round are independent of loss in other rounds (*TCP* [43]).

Some models work very hard to generalize the assumptions that are necessary for the derivations. Such efforts are unnecessary—it serves no purpose to precisely determine the assumptions (e.g., Markov regenerative process) that are needed to justify an equality when so much else in the model are approximations. They only transfer the need for justification from the equation to the assumptions. One should just regard the equality as another approximation, and let the experimental measurements justify the derivation.

Analytical models are often dismissed as unrealistic because the assumptions are simplistic. However, the results from analytical models are often robust with respect to violations of its assumptions; in other words, the assumptions are just a means to an end, a way of pressing ahead with the analysis, and may in fact be stronger than necessary. One should look pass the assumptions and approximations to see what the model delivers, and wait for the experimental validation.

It follows that we should also not hesitate to apply a theoretical result even if its derivation is based on an assumption that is violated. For example, the MVA algorithm in Section 5.3 assumes service demands are constant. This does not prevent some models from using the algorithm even when service demands change with the queue size *PerformanceAssurance* [45] or the number of jobs *DatabaseScalability* [14]. Here, one can view the underlying Arrival Theorem as an approximation for the system.

### 10.1.4  METRICS

For a given system, the appropriate performance measures to focus on are usually obvious. However, there are often secondary metrics that arise naturally in the model, but are distractions. In transaction locking, for example, the main performance measures are throughput and response time, but any model must first estimate the probability that a lock request encounters a conflict. This probability $P_{\text{conflict}}$ is, in fact, not interesting by itself: Transactions that require many locks may have a small $P_{\text{conflict}}$, yet have a high probability of being blocked at some point in its execution; moreover, a small error in estimating $P_{\text{conflict}}$ can translate into a large error in estimating throughput, since that small error is repeated for each lock request.

For another example of why one should always keep an eye on the big picture, consider $P_{\text{deadlock}}$, the probability that a lock request results in a deadlock. Although $P_{\text{deadlock}}$ starts out smaller than $P_{\text{conflict}}$, the deadlock probability grows at a faster rate as workload (e.g., number of concurrent transactions) increases. Thus, one may observe some phenomenon, measure $P_{\text{deadlock}}$ and $P_{\text{conflict}}$, and correctly conclude that the phenomenon is caused by deadlocks. However, by the time deadlocks have a significant impact on performance, the system is already in its performance degradation zone ($k^2 N/D > c$), so one is in fact studying a phenomenon that is of questionable interest.

### 10.1.5  SCIENCE AND TECHNOLOGY

The impact of transaction locking on performance is the result of an intricate interaction between data and resource contention: Intense data contention can cause many transactions to be blocked, thereby leading to more context-switching and swapping, both of which add to resource contention. On the other hand, if more transactions are blocked, then there is less demand on processor cycles and disk bandwidth, so resource contention is reduced. How should one study this interaction?

One could point to the complexity of the mutual dependency as a justification for using only an experimental or simulation study. However, the generality of the conclusions from such a study would be limited by the particular hardware/software implementation or the simulation parameters.

For performance analysis to be a science, its results must withstand changes in technology. For transaction locking, we seek an analysis of data contention that is independent of the hardware and software underlying the system, yet permits us to draw conclusions about transaction performance with different levels and modes of resource contention.

Such an analysis can be done by modeling resource contention with a variable $T$, the time between two lock requests if the transaction is run alone (i.e., without dilation from blocking). If we consider this variable $T$ to be a function of the number of active (i.e., nonblocked) transactions $N_{active}$, then data contention determines $N_{active}$, while resource contention determines $T(N_{active})$.

Now, $N_{active}$ is a random variable, so evaluating the data contention to determine $N_{active}$ may involve a recursion that incorporates both data and resource contention at other values of $N_{active}$ [38]. However, if we use just the average value of $N_{active}$ to estimate $T(N_{active})$, then it is possible to evaluate the data contention and resource contention independently (e.g., a Markov chain for the former and a queueing network for the latter), then integrate the two [56]. We see here that Average Value Approximation (AVA) is useful not just for derivations, but it is also an analytic technique for decoupling two closely interacting forces in a system. Such a decoupling provides a framework for reasoning informally to an engineer about system behavior. A nagging worry, however, is that some important synergistic effect may be lost through this decoupling.

## 10.1.6   INTUITION AND CONTRADICTION

To design and engineer a complicated system, we need help from intuition that is distilled from experience. Experience with real systems is limited by availability and configuration history. We can get around that through simulation experiments. However, the overwhelming size of the parameter space usually means we need to limit its exploration, but experiment design—and its interpretation—may, in turn, be swayed by intuition. One way of breaking this circularity is to construct an analytical model that abstracts away the technical details and zooms in on the factors affecting the central issues. We can then check our intuition with an analysis of the model.

Intuition is improved through contradictions: they point out the limits on the old intu-ition, and new intuition is gained to replace the contradictions. Again, such contradictions can be hard to find in a large simulation space, but may be plain to see with an analytical model.

In the case of transaction locking, there were several examples of old intuition being con-tradicted. For example, one very simple conflict resolution policy would be to restart a transaction if a lock it needs is already held by another transaction. This is the so-called *no-waiting* policy. Intuition suggests that this policy will perform badly: why restart a transaction (thus losing what it has done so far) when it costs nothing to just block it? In fact, there *is* a cost in blocking. When a transaction is blocked, it is not "using" the locks that it already holds, while preventing other transactions that need those locks from getting them and making progress. Even if one could immediately see this anti-social effect of blocking, it is still difficult to believe that it could be worse than the draconian no-waiting policy. But it could.

For another example, recall that static locking is where a transaction does not begin ex-ecution until it gets all the locks it needs. In contrast, dynamic locking is where a transaction gets a lock only when it needs the lock. In both cases, the locks are held till the end of the transaction. Intuitively, static locking is a loser, since the transaction is getting locks before it needs them, thus reducing the level of concurrency. This reasoning turns out to be superficial. There are conditions under which locks are held longer under dynamic locking than under static locking.

The third example concerns throughput degradation as concurrency increases. Intuition may tell us that this is caused by deadlocks and restarts, i.e., the system has been driven to a point where it is spending too much of its time redoing deadlocked transactions, thus losing its throughput. Such an intuition would be consistent with page thrashing in operating systems, but is wrong.

These three examples are not hypothetical—they can be found in the literature. How could one's intuition go wrong?

In the example of the no-waiting policy, there are at least two reasons why the intuition is partly correct. One of them is the confusion between resource and data contention, and the other involves looking at the appropriate performance measure. An excessive amount of restarts is bad only because it causes intense contention for resources; as a means of resolving data contention, restarts are no worse than blocking in their effect on performance, if we factor out the resource contention. Also, restarts are bad for response time, but not necessarily bad for throughput. In a sense, restarts allow the system to pick from the input stream those transactions that can get through with no hassle, and delay the rest till the conflicting transactions have left.

In the second example, there is a critical difference between static and dynamic locking: a transaction holds no locks when it is blocked under static locking, while a blocked transaction under dynamic locking may be holding locks. Each time a transaction is blocked under dynamic locking, the period over which it holds its locks is lengthened. It is thus possible that transactions end up holding on to locks longer than if they had used static locking, so it is not clear that static locking is really a loser.

In the third example, locking can cause throughput degradation not through deadlocks, but when too many transactions are tied up waiting for locks that they need but cannot get. For each added transaction, more than one transaction may become blocked, so throughput drops. The picture of frenzied activity for page thrashing in operating systems leads one to expect a similar picture for database systems, when in fact the appropriate one is of a large number of idling transactions waiting for a few transactions to finish, so they can proceed.

The failure of intuition in these examples is partly caused by an unfamiliarity with the impact of data contention and its interaction with resource contention. To give another example, recall the issue of lock granularity. The intuition there is that covering more data per lock will reduce the number of concurrent transactions but save on instruction overhead. Now, refining granularity increases the number of data items $D$ as well as the number of locks $k$ that a transaction needs. Once an analytical model shows that performance is determined by $k^2 N/D$, it becomes clear from $k^2$ that an increase in $D$ can in fact reduce concurrency. Moreover, an increase in concurrency can worsen performance because concurrent transactions compete for resources.

Incidentally, the first and second examples (on no waiting and static locking) concern extreme policies. Other extreme policies can be found in cache management, admission control, load balancing, etc., and they are often believed to give worst-case performance. The no-waiting policy and static locking above show that such a belief can be wrong.

The ability to reveal and resolve contradictions, and thus improve engineering intuition, is one demonstration of the power of analytic modeling.

## 10.2   DISCUSSION OF PAPERS

Our final discussion includes another four papers: *802.11* [55], *SoftState* [35], *CloudTransactions* [31], and *ElasticScaling* [11].

*StreamJoins* [28]
> The authors made good use of the model to analyze the issues that motivated the paper. In *Sec. 5.1*, for example, they used the equations to derive closed-form expressions for when one particular join combination is better than another. These expressions in *Eq. (8)* and *Eq. (11)* would be more informative if the parameters $B/|B|$, $I_h$, $P_h$ etc. were not instan-

tiated, so one could better understand how the comparison depends on these parametric values.

*SleepingDisks* [68]

This paper uses the queueing model not for analysis, but for the calculation to minimize energy consumption subject to the response time constraint $R_{limit}$ (*Sec. 3.1.1*).

*GPRS* [42]

*Sec. 5* uses the analytical model to plot performance curves for analysis. This is an example of analytic simulation: those curves could have been plotted with the simulator, with greater fidelity.

It turns out that the performance metrics ($p(n)$, $\tilde{U}$, etc.) depend on $t_B$, $t_{off}$, $x_B$ and $x_{on}$ only through $x = \frac{t_B x_{on}}{x_B t_{off}}$ in *Eq. (6)*. This is an example of how the parameter space is reduced with the help of an analytical model.

For example, *Sec. 6* uses the model for dimensioning: Given a maximum acceptable blocking probability (0.02, say), what is the maximum number of cells, or maximum number of mobiles per cell? The answer can be found from *Fig. 14* and *Fig. 15*, in terms of $x$. Without the introduction of $x$ from the analytical model, a simulation study (say) of dimensioning using $t_B$, $x_{on}$, etc. would be quite intractable.

*Fig. 12* shows that, for a fixed number of users per cell and as $x$ increases, throughput first increases, then decreases. Such nonmonotonic behavior is more interesting than the monotonic curves in previous plots, and bears closer study. For example, it suggests that $x$ should only increase till where throughput begins to decrease, thus further delimiting the parameter space.

*InternetServices* [63]

Based on the queueing network model of the system, the MVA algorithm is used for dynamic provisioning, to determine the number of servers needed to maintain a response time target (*Fig. 12b*) as arrivals change over time (*Fig. 12a*).

*TCP* [43]

Notice that *Eq. (31)* expresses TCP throughput $B(p)$ in terms of loss probability $p$ and round trip time RTT. Clearly, for any nontrivial Internet path, $p$ and RTT can only be measured, not predicted. If one is to measure $p$ and RTT, one might as well measure $B(p)$ too, so what is the point of having that equation?

The significance of the equation lies not in predicting $B(p)$, but in characterizing its relationship with $p$ and RTT. Such a characterization led to the concept of *TCP-friendliness* and the design of *equation-based protocols* [17].

*CodeRed* [69]

The parameter tuning needed for *Eq. (17)* to fit the observed infection means that the model cannot be used to predict the epidemic at the beginning of the infection. However, knowing that the equation does incorporate the major factors shaping the infection, one can use it to analyze the effectiveness of a specific countermeasure. It may even be possible to calibrate the parameters dynamically (as data is received), to help choose the countermeasure for curbing the spread.

*MediaStreaming* [62]

The authors use their analytical model to show that the multiplexing protocol in *Theorem 3.1* is optimal. Such a proof of optimality is hard to do by simulation.

As *Fig. 7a* shows, if the failure rate is sufficiently high, then the system may not be able to generate enough peers to support the demand (so rejection rate never reaches 0); the model does not say how big $s$ can get before this happens.

*TransactionalMemory* [24]

The model is used for analytic simulation, to compute *Table II* and *Table III*, and thus draw conclusions about the performance of software transaction memory. It is likely that much more insight can be drawn from this model, as demonstrated in the no-waiting case for database transactions [58]. The key lies in simplifying the expression for $q_i$ in *Eq. (4)*. This is the issue, discussed in Section 9.2.1, of how a transaction's locks affects its own performance.

*SensorNet* [49]

This paper gives a good example of how an analytical model that provides closed-form expressions (like *Eq. (18)*) can bring crisp insight to a problem.

However, the key performance metric is data success rate $S$. It is usually difficult to obtain explicit, closed-form expressions for such a probability, so there is little that one can do with the formulas for $S$ in *Sec. VIII* besides plotting them.

*NoC* [29]

There is no demonstration of how the analytical model is used to evaluate their router design.

*DatabaseSerializability* [3]

This paper is an example where the queueing network is used as a simulation model and not as an analytical model for deriving equations to describe the system. The queueing network is not separable; e.g., *Fig. 3* shows no "servers" for the Ready Queue and Blocked Queue.

Note from *Fig. 6* that the data for Strict2PL reaches a peak, then decreases as multiprogramming level increases. (This may also happen for the BSAFC curves for higher write probabilities.) This subsequent decrease is not caused by resource contention, since there are infinite resource units. Rather, it is the result of data contention—see Sec. 10.1.6.

We see here that whether "infinite" resources is realistic is beside the point, as it is a way of factoring out resource contention so we can clearly see the effect of data contention. Also, one must not assume that "infinite" resources give a performance bound; if more caches can cause more data contention, "infinite" resources may result in more data contention and poorer performance than "finite" resources.

One can argue that freshness constraints bring a third force into concurrency control performance, namely, *time contention* [52]. Intuitively, resource contention increases if more transactions want to use a resource (CPU cycle or disk access, etc.), and data contention increases if more transactions require locks (on records or tables, etc.). Similarly, time contention increases if more transactions need to satisfy their freshness constraint: more reads are rejected and transactions are aborted. One may object that this is just another form of data contention or resource contention. On the other hand, one could also consider data contention as just another form of resource contention (the data lock being the resource), but it is now generally accepted that the two are different.

Any analytical model that can establish or dismiss time contention as a third force would be making a scientific contribution that goes beyond the technological issues of cache deployment, serialization algorithm, etc. To do so, such a model would need to provide closed-form expressions for analysis; a queueing network (like the one in this paper) would not suffice.

A comparison of *Fig. 5* and *Fig. 6* shows that Strict2PL is above BSAFC-0 in one plot, but below in the other. They illustrate how two simulation studies of the same protocols can reach different conclusions if the parameter values are set differently. This is why a performance study must carefully delimit and explore the parameter space, and why the use of "magic constants" (like 1 s for think time in *Table 1*) is an issue.

Furthermore, the performance reversal between Strict2PL and BSAFC-0 is not determined by data contention, but by resources contention (since the difference between *Fig. 5* and *Fig. 6* lies in the resource contention).

#### *DistributedProtocols* [20]

The analysis in *Sec. 4.1.3* shows that the endemic protocol has liveness (since $\gamma > 0$), fairness (since the protocol is symmetric) and safety for the second equilibrium. One can view the space-time plot in *Fig. 8* as an analytic validation of those claims.

This paper gives an excellent demonstration of the power of an analytical model: it shows how such a model can be used to design protocols.

#### *WebCrawler* [8]

The design of the crawler was based on scientific results derived with the simple model; e.g., batch and incremental crawling yields the same average freshness, and the optimal crawl rate defines a threshold on the update frequency.

*DatacenterAMP* [22]

The technology for marshalling the resources of multiple small cores into one large core is not ready. This paper is therefore a theoretical study that demonstrates the power of using a model to analyze what is (or is not) possible.

The plots (*Figs. 4, 5, 6*) are generated by a numerical evaluation of the equations, and conclusions drawn from looking at these plots. The authors should have analyzed the equations instead.

*RouterBuffer* [1]

The model shows that (i) for $n$ long flows, buffer size should be inversely proportional to $\sqrt{n}$ and (ii) for short flows, $\mathrm{Prob}(Q \geq B)$ does not directly depend on $\overline{RTT}$, $C$, nor $n$. These are insights obtained through the analytical model—one cannot get them from any finite number of simulation experiments.

*DependabilitySecurity* [61]

This paper is an excellent illustration of how powerful Markov chains can be. They are highly flexible, and can even be used to quantitatively model concepts like confidentiality and dependability.

However, they can provide closed-form expressions only for simple cases, like the 1D birth-death process in Chapter 3. Most of the time, their solution is only implicitly defined by some set of equations that are to be evaluated numerically and iteratively.

*PipelineParallelism* [41]

Like for Markov chains, the solution to a queueing network is usually defined implicitly by some set of equations. For example, the authors use the work stealing model to calculate the performance of the TBB implementation and compare that to the Pthread implementation (*Figs. 10, 11*).

However, in some special cases, it is possible to extract closed-form expressions from a queueing network model. The work stealing model here was previously formulated for a multimedia system [57], where special features of the system could be exploited to extract closed-form expressions for sizing the number of multimedia servers, link, and switch bandwidth.

*Roofline* [66]

The roofline model is a simple bottleneck analysis, yet it is sufficiently powerful for analyzing the issues that the authors raised: (a) To know if a kernel's performance is constrained by the architecture's memory bandwidth or processor speed, one simply compares its operational intensity to the ridge point. (b) To choose an architecture for a kernel, one compares their rooflines. (c) To evaluate the efficacy of a kernel (or compiler) optimization, one considers how the optimization shifts the roofline and changes operational intensity.

*Gossip* [2]

In the experimental validation of *Fig. 4*, the model calculation is itself an analytic simulation, i.e., using the equations for $P(01|10)$, etc., to simulate the data exchange. How persuasive the comparison is then depends on how similar the model simulation is to the protocol simulation. For example, is the data exchange synchronized in exactly the same way for both simulations?

One can see the power of an analytical model from the conclusion that the optimal shuffle size is $s = n - \sqrt{n(n-c)}$. It shows that the optimal $s$ does not depend on the number of nodes $N$; also, if the number of items $n$ is large, then the optimal $s$ is approximately $c/2$. Moreover, $\frac{ds}{dn} < 0$, so the optimal $s$ decreases with $n$. It is hard to see how one can get such insights into the protocol by simulating it.

Note that the explicit form of *Eq. (1)* is not enough for analyzing the protocol. In particular, the derivation of $s = n - \sqrt{n(n-c)}$ depends crucially on the use of AVA in *Sec. 3.3*, so one can easily determine the $s$-derivative. In contrast, finding the optimum $s$ by simulation would require a tedious number of experiments.

Instead of trying to justify that approximation for $P_{drop}$ (which is hard to do rigorously), one should simply validate the results it leads to.

*EpidemicRouting* [67]

This model is clearly not for quantitative prediction since that requires a value for $\beta$. Rather, the power of this fluid model is evident from the results in *Table 1* for three variations of the epidemic protocol. Stochastic models can rarely provide such closed-form expressions to show explicitly how the input parameters affect the performance measures. One can, for example, use these expressions to determine the conditions under which one protocol is better than the other (and then verify the conclusion with experiments).

However, a theoretical analysis of the tradeoff between two metrics is difficult even with the closed-form expressions, so *Sec. 4* uses them for a numerical analysis, i.e., for analytic simulation.

*P2PVoD* [16]

The authors observe that, contrary to intuition, a reduction in transfer sequentiality can increase the throughput. *Sec. 5.1* demonstrates this with the hybrid scheme that is controlled by a parameter $s$ for sequentiality ($s \approx S^{hybrid}$) and considering how $s$ affects $T^{hybrid}$. This is an excellent example of how an analytical model can reveal and explain counter-intuitive behavior in a system. It also illustrates how the model reflects the tradeoff when one moves from one point to another in the design space.

The Tradeoff Theorem in *Sec. 4.3* is a good example of extracting a scientific result from a performance model. However, one should bear in mind its underlying assumptions, such as the independence of bad peers (used in the definition of robustness). The paper should present experiments to show that this result is robust with respect to violations of its assumptions.

*SoftErrors* [40]

The authors call their model "mechanistic," in the sense that they use equations to model details in the dynamics of the microarchitecture. Such a model is also called **whitebox**, in contrast to **blackbox** models (like regression in statistics or neural networks in artificial intelligence). A whitebox model can be used to analyze the factors driving system behavior, since its variables and equations capture details of what goes on in the "box;" in contrast, a blackbox model offers no such insight into what goes on in the "box."

By "first-order," the authors mean that they model separately the events (e.g., cache misses and mispredictions) that have highest significance. Modeling an interaction between two such events (e.g., *Sec. 3.1.5*) might be called a "second-order" analysis.

A larger reorder buffer or dispatch width can improve performance, but also increase the vulnerability. The AVF model is used to study this tradeoff, via an analytic simulation to generate *Figs. 6* and *8*. Here, design space exploration by simulation would take too long, since soft errors are rare and their impact is transient. Moreover, the model facilitates a separation of the contributions from cache misses, mispredictions, etc., and *Eq. (3)* does help explain the difference between *gobmk* and *namd* (*Sec. 5.1*).

In formulating their model, the authors were careful to balance simplicity, numerical accuracy, insight and data availability. They invoke previous work to support the assumption that cache misses and mispredictions divide the timeline into *independent* intervals.

One might expect a workload that has a low CPI to have a low AVF, but *Sec. 5.3* points out that this is contradicted by *namd*. This intuition is clarified in *Sec. 5.1* with the help of *Eq. (3)*.

The separation of critical path and latency profiling is a way to decouple the workload (which determines the critical path) and the microarchitecture (which determines the latency), so one can evaluate the impact of a change in hardware (reorder buffer or dispatch width, say) without a need to re-profile the workload. This is similar to decoupling data and resource contention in database locking (Sec. 10.1.5), and is a helpful technique for a scientific analysis of a complicated system.

*GPU* [26]

The flexibility of a Markov chain makes it a popular modeling technique, but one must remember that it is memoryless. *Figs. 11* and *12* show that this weakness makes it no better than a naive method for modeling warp scheduling.

*Eq. (21)* puts a bound on the bandwidth delay. One should not worry about such a violation of the $M/D/1$ model, since the queueing model is itself an approximation that has a serious artifact, namely unbounded delays.

*Sec. VI-C* demonstrates how the model can explore the impact of changes in DRAM bandwidth, MSHR entries and warps per core—there is no practical way of doing such exploration with real hardware. It can also generate CPIstacks, similar to the AVF model in *SoftErrors* [40] breaking down the contribution from different structures in the microarchitecture.

In *Sec. VII*, the authors point out that it is *counter-intuitive* that `kmeans_invert_mapping` spends many cycles (`QUEUE`) queueing for DRAM bandwidth, yet have negligible stall cycles (`DRAM`) from L2 misses. Their explanation for this apparent contradiction is a refinement of the intuition that is gained from the CPIstack that is constructed by the model.

### *Database Scalability* [14]

Most of the plots in the paper can be obtained with bottleneck analysis. This may suffice if we only want to know the bound on throughput, and how response time scales as the number of replicas increases. Such an analysis is very simple, but it already says that, for running TPC-W on a single-master system, we should not go beyond 4 replicas (*Fig. 8*). This is nontrivial insight into the system's scalability, considering that we only have standalone profiling in hand.

Note that $p = 1/DbUpdateSize$ assumes the writes are uniformly distributed. This assumption can be relaxed; e.g., see the use of Application Contention Factor in *ElasticScaling* [11].

### *Performance Assurance* [45]

*Fig. 12* shows that CPU utilization is nonmonotonic with respect to the number of clients. Such behavior cannot be modeled by traditional separable queueing networks. MAQ-PRO is able to capture this monotonicity because it modifies the service demand to model the impact of context switching, etc., (*Fig. 5*).

The model is a multiclass queueing network where a job class is a (RUBiS) service and each job is a thread. Standard MVA requires first-come-first-served queueing and all classes to have the same average service time per visit; these assumptions may be fine for the disks in *Fig. 1*, but perhaps not for the CPU.

Thus, in addition to being ready to adopt strong assumptions to facilitate derivations, we should also not let violation of assumptions (like separability) hold us back from a solution technique (like MVA) that is derived with those assumptions. In both cases, we should let the experimental validation decide if the resulting model is acceptable.

### *Caching Systems* [36]

*Fig. 5* is presented as a validation, using Youtube videos, of the insights from the analysis. However, these insights are not obtained from a mathematical analysis of the model, but from looking at the plots generated with the model. Since these plots could just as well be obtained from the simulator, this is an example of analytic simulation.

Instead, the authors should have focused on the small cache approximations in *Sec. IV-G* because (i) caches are not interesting if they are large and (ii) closed-form expressions for hit probability are very hard to derive, and offer an excellent tool for analyzing the differences among replacement policies. Memory is a fundamental resource for computing, and caches will be around no matter how technology changes, so a model that advances our understanding of cache behavior would be a contribution to the *science* of computing.

*InformationDiffusion* [37]

Although the pieces of a system are artifically created, they can exhibit a hard-to-understand, organic behavior when put together. This paper illustrates how information diffusion can be analyzed like epidemiology in the biological sciences.

*MapReduce* [65]

*Fig. 8* shows that performance has hit a bottleneck, but utilization is still far from 100%. The bottleneck therefore lies not in queueing for CPUs and disks, but in the task synchronization. This paper missed the opportunity to use the model to study how these two factors—resource contention and precedence constraint—interact to determine MapReduce performance (analogous to how resource contention and data contention interact to determine transaction performance in a database system).

*802.11* [55]

The system consists of a wireless cell with stationary mobile nodes using IEEE 802.11 (basic access) to send traffic to the access point. Packet collisions induce retransmissions and backoff, so maximum possible throughput is lower than the channel bandwidth. The issue is how this saturation throughput depends on the parameters, and on the tradeoff between collision and backoff.

The main parameters are the number of nodes $n$ and the minimum contention window size $W$; the main performance measures are the collision probability $p$ and the saturation throughput $S$. *Table 1* lists other parameters and metrics.

*Fig. 2* shows the model's key idea, formulated—using AVA—as $p = 1/W_{\text{backoff}}$. The rest of the model follows in a straightforward manner from the first-order approximation in *Eq. (4)*. It is possible that such an approximation may result in a trivial linear model, but the experiments show that it in fact suffices to model the nonlinear behavior of the protocol.

The parameter space is first restricted to where $p < 0.5$. A comparison of *Figs. 3* and *4* shows that there is throughput saturation even under such a restriction. This is similar to the case for transaction locking, where throughput begins to degrade even before the deadlock probability becomes significant.

The space is further reduced through three results: *Claim 1* reduces the parameters to $W$, $m$ and $n$; *Claim 2* to $m$ and $g(= W/(n-1))$; and *Claim 3* to just $g$.

Given the multiple approximations, some gross (like *Eq. (12)*) and some arbitrary (like *Eq. (13)*), there is a clear need for analytic validation of the claims. This is done in *Table 3* (for *Claim 1*), *Figs. 7* and *8* (for *Claim 2*), and *Table 5* (for *Claim 3*). Note that these tables and figures are generated by the simulator alone, to validate the claims generated by the model.

There is further analytic validation in *Table 4* (for *Cor. 1*), *Table 6* (for *Claim 6*), and *Table 7* (for *Claim 7*).

The insights gained from analyzing the equations are captured in *Cor. 1* (the equivalence between minimum window size $W$ and number of nodes $n$), *Cor. 2* (how $W$ should vary with $n$ and packet size), *Claim 2* (how the gap $g$ transforms *Fig. 3* into *Fig. 7* and *Fig. 4* into *Fig. 8*), and *Claim 6* (how maximum throughput is a tradeoff between collisions and backoffs). Such insights are hard to get if the analytic technique does not yield closed-form expressions. For example, a typical Markov chain model for 802.11 would have one dimension for $W$ and another for $n$, so one cannot see from such a 2-dimensional model that performance actually depends on $W/(n-1)$, without already knowing what to look for.

*SoftState* [35]

The use of soft states is a paradigm favored by many designers of distributed protocols. This paper provides examples of how the hardware and software can be abstracted away by a model to provide a scientific basis for engineering intuition.

The system consists of two parties running some connection protocol over a possibly lossy channel, with attention focused on the exchange of control messages. The central issue is the difference between hard and soft state, which is characterized as a tradeoff between the cost of refreshing state and the cost of stale states.

The model used varies with the setting: a flow diagram for protocol state (*Fig. 2*) in *Sec. 3*, a 2-dimensional Markov chain in *Sec. 4*, and an $M/M/\infty$ server in *Sec. 5*. Compare the insight gained from the models in *Sec. 3* and *Sec. 4*: the simple flow diagram results in a concise *Eq. (1)* that characterizes the tradeoff between hard and soft state, in terms of connection lifetime, loss probabilities and costs; in contrast, one cannot extract any closed-form expression from the Markov chain, since its solution is only implicitly defined. Note the state aggregation (Sec. 4.3) that reduces the $N$-dimensional state space to a 2-dimensional $\mathcal{M}_h$.

Although the analysis starts off by deriving a closed-form for the optimal refresh interval (*Eq. (1)*), the rest of the paper uses the models for analytic simulation (i.e., plotting graphs). Ideally, there should be some analytic validation to verify the conclusions from such an abstract study; e.g., check if measurements with real protocols running over some real network can reproduce a plot like *Fig. 4(a)*. Such validation is necessary, to guard against behavioral artifacts that may result from the models (e.g., *Eq. (2)* for channel loss).

*Fig. 3(a)* has a logarithmic horizontal axis, and a vertical axis that does not start from 0; these tend to visually exaggerate the difference between the two curves. Also, the relative positions of the two lines depend on the values of $a$, $b$, $C_{ss}$ etc.

*CloudTransactions* [31]

This paper does not have a performance model. The purpose of including it in our discussion is to demonstrate how even a simple analytical model can guide the experiments and provide insight into its results.

Cloud providers use elasticity as one selling point for their services. these providers offer a choice of architectures (*Figs. 1–5*), and this paper's objective is to compare their scalability and cost for a transaction processing workload.

The transactions are from the synthetic TPC-W benchmark of 14 request types for an online bookstore. This workload is chosen to exercise the software stack of web, application and database servers, with storage that is distributed.

The database size is fixed at 10,000 items that add up to 315 MB of raw data and 1 GB of indexes. It is not clear if the results for such a small dataset can be extrapolated to sizes that have to be orders of magnitude larger to sufficiently stretch the cloud architectures. The control parameter in the experiments is the number of emulated browsers (EBs), where each EB simulates one shopper who issues multiple requests, one at a time (i.e., the system is closed). In the experiments, EB does not go beyond 9000. This magic value affects the conclusions, as we shall see.

The authors use some cost-related metrics to compare the architectures but, from the scientific viewpoint, there is nothing much that one can say about the cost metrics. As the authors observed, "The cost analysis is more an artifact of the business model."

The key metric is web interactions per second (WIPS). WIPS includes a request only if it satisfies its response time bound and is thus considered valid, i.e., WIPS is a measure of **goodput**. *Figs. 8–10* show that, as EB increases, the goodput (green line) and issued requests (yellow line) eventually diverge. The issued requests represent input rate $\lambda_{in}$, and goodput represents (valid) output rate $\lambda_{out}$. The divergence between $\lambda_{in}$ and $\lambda_{out}$ here indicates that the system is not in steady state (see Chapter 8). Note that, using the notation of Fig. 8.2b, the WIPS graphs are plotted against $N$ (EB), not $K$.

The experimental comparison focuses on throughput limits, so we can use bottleneck analysis (Chapter 6) to study the asymptotic bound as EB increases. Note that bottleneck analysis cannot model the throughput degradation seen in **partition+replicated** architectures (SDB and AppEng)—modeling such degradation would require an analysis of the consistency and replication protocols. The table below lists the variables in our analysis.

| Model Parameters | |
|---|---|
| $N_{WA}$ | Number of web/application servers *without* co-located database servers |
| $N'_{WA}$ | Number of web/application servers *with* co-located database servers |
| $N_{st}$ | Number of storage servers *without* co-located database servers |
| $N'_{st}$ | Number of storage servers *with* co-located database servers |
| $D_{WA}$ | Service demand (seconds on $\mathcal{M}$) per transaction at web/application server |
| $D_{st}$ | Service demand (seconds on $\mathcal{M}$) per transaction at storage server |
| $D_{db}$ | Service demand (seconds on $\mathcal{M}$) per transaction at database server |
| $c$ | Ratio of database server speed in **classic** architecture to $\mathcal{M}$ |
| $\lambda$ | Transaction arrival rate |

We make three assumptions: **(A1)** Web/application servers and storage servers use similar (commodity) machines $\mathcal{M}$. **(A2)** For the **classic** architecture, the database server uses a machine that is $c$ times as fast as $\mathcal{M}$. **(A3)** Transaction arrivals are evenly distributed among the servers.

For **classic**, (A3) implies that the transaction arrival rate at a web/application server is $\lambda/N_{WA}$. In steady state, the server utilization $\lambda D_{WA}/N_{WA}$ is bounded by 1, so we get an asymptotic bound $\lambda \leq N_{WA}/D_{WA}$. Similarly, $\lambda \leq N_{st}/D_{st}$. By (A2), the service demand at the database server is $D_{db}/c$, so $\lambda \leq c/D_{db}$. Together, we get:

$$\lambda \leq \lambda_{classic} = \min\left\{\frac{N_{WA}}{D_{WA}}, \frac{c}{D_{db}}, \frac{N_{st}}{D_{st}}\right\}. \tag{10.1}$$

Similarly, we get, for **partition**,

$$\lambda \leq \lambda_{partition} = \min\left\{\frac{N_{WA}}{D_{WA}}, \frac{N'_{st}}{D_{db} + D_{st}}\right\} \tag{10.2}$$

and, for **distributed control**,

$$\lambda \leq \lambda_{dist.control} = \min\left\{\frac{N'_{WA}}{D_{db} + D_{WA}}, \frac{N_{st}}{D_{st}}\right\}. \tag{10.3}$$

*Fig. 7* shows that the two **partition+replicated** architectures (AE/C and SDB) perform worse than **classic** (MySQL and RDS). Is this inherent in the architectures? *Sec. 6.1* notes that, for these architectures, the database server is the bottleneck, so

$$\lambda_{classic} = \frac{c}{D_{db}} \quad \text{and} \quad \lambda_{partition} = \frac{N'_{st}}{D_{db} + D_{st}}.$$

Therefore, $\lambda_{partition}$ can be raised to match $\lambda_{classic}$ in *Fig. 5* if

$$\frac{N'_{st}}{D_{db} + D_{st}} = \frac{c}{D_{db}}, \quad \text{i.e.,} \quad N'_{st} = \left(1 + \frac{D_{st}}{D_{db}}\right)c.$$

In other words, the **partition+replicated** architecture does not have a throughput that is inherently less than **classic**; rather, its throughput can be scaled up by adding more database/storage servers.

*Fig. 7* also shows that **distributed control** (S3) appears to scale linearly as EB increases. Since no system scales linearly forever, where is the limit for **distributed control**? For the configuration in the experiments, the authors noted that a medium EC2 machine can support 1500 EBs as web/application server and 900 EBs as web/application/database server. This suggests $1500 D_{WA} = 900(D_{WA} + D_{db})$, i.e., $D_{WA} = 3D_{db}/2$. Moreover, $N_{WA} = 6$ and $N'_{WA} = 10$. Therefore,

$$\lambda_{classic} = \min\left\{\frac{4}{D_{db}}, \frac{c}{D_{db}}, \frac{N_{st}}{D_{st}}\right\} \quad \text{and} \quad \lambda_{dist.control} = \min\left\{\frac{4}{D_{db}}, \frac{N_{st}}{D_{st}}\right\}.$$

*Fig. 7* shows $\lambda_{classic} \neq \lambda_{dist.control}$, so we must have $\lambda_{classic} = \frac{c}{D_{db}}$. Therefore, $\lambda_{classic}$ can be raised to match $\lambda_{dist.control}$ by increasing $c$, i.e., speeding up the database server.

In particular, $\lambda_{classic} = \lambda_{dist.control}$ if $c \geq 4$. Moreover, if $\frac{N_{st}}{D_{st}} \geq \frac{4}{D_{db}}$, then $\lambda_{dist.control} = \frac{4}{D_{db}} = \frac{4}{c}\lambda_{classic}$; this says that, if $c = 1$ in the experiments, then the **distributed control** architecture (S3) in fact saturates at about 1800 WIPS. This would happen at about 13,500 EBs.

A topic like cloud architectures may seem to be all about technology, but we again see that one can extract some science with a performance model: The simple bottleneck analysis provides insight into the effect of $N'_{st}$ for **partition** architectures and the effect of $c$ for **classic** architectures. It can also guide the design of the experiments, e.g., how far to push EB in order to see saturation for **distributed control** architectures.

The magic number of 9000 EBs therefore prevents the experiments from revealing a crucial limit to the architecture.

*ElasticScaling* [11]

One argument that service providers offer for hosting a system in the cloud is that resources can be scaled elastically to respond to demand. This scaling can be done more effectively if there is a tool for exploring the parameter space, e.g., increasing or decreasing the number of machines. For a tenant, developing such a tool is harder since they have very little knowledge of the system architecture.

TAS illustrates how this issue can be addressed in the case of distributed transactional memory, where there is data contention in addition to the resource contention. It has a *whitebox* that uses equations to model the data conflicts and resulting delays, and a *blackbox* that uses machine learning to predict the delays from resource contention. The two models get inputs from each other (*Fig. 2*)—like in the hierarchical decomposition in *MapReduce* [65]—so TAS iterates between them to derive a fixed-point solution.

Blackbox models are seductive: if machine learning can predict delays without knowing the details of the architecture, why not let it deal with the data contention as well? *Fig. 6* shows that such a model may have a serious artifact: making linear predictions when system performance is, in fact, nonlinear.

*Fig. 5* shows that metrics calculated with TAS match real measurements for commit probabilities that are as small as 0.6. While impressive, this is possibly wasted effort on a part of parameter space that has unacceptable performance. If, in practice, one requires the abort probability to be small, then that requirement can be used to simplify the expressions in *Sec. 4.1.2* and thus derive closed-form formulas for the metrics.

Perhaps the hardest part of modeling transaction performance lies in estimating the probability of a conflict. Most papers typically assume requests are uniformly distributed over the data items. This is the case for *Eq. (1)*, which uses Little's Law to derive the lock utilization $\lambda_{lock}T_H$. If this utilization is used to estimate $P_{abort}$, then that is a PASTA approximation (similar to $p_{in}(m) = p_{hit}(m)$ for *CachingSystems* [36], since the lock requests may not be Poisson.

The "mean-value analysis" mentioned in *Sec. 4.1* does not refer to the MVA algorithm, but to AVA. For example, request rate and lock holding time can vary from item to item, so $\lambda_{lock}$ and $T_H$ are averages over the data items. Therefore, for nonuniform access, $\lambda_{lock}T_H$ is an AVA estimate for the total utilization (which can exceed 1, like in the case of a multiserver queue).

One possible model for nonuniform requests is $b - c$ *access*, where a fraction $c$ of data items are *hot*, a fraction $b$ of requests are for hot items, and requests are uniformly distributed over the hot items, and similarly over the non-hot items. Under certain assumptions, one can prove that $b - c$ access over $D$ items has the same conflict probability as uniform access over $D/\alpha$ items, where $\alpha = 1 + \frac{(b-c)^2}{c(1-c)}$. In other words, $b - c$ access over $D$ items is equivalent to uniform access over as *smaller* set of $D/\alpha$ items [56].

It is hard to cleanly generalize that result to other nonuniform distributions. Instead, TAS uses *Eq. (2)* to empirically define ACF for *any* nonuniform access distribution, thus making nonuniform access equivalent to uniform access over $\frac{1}{ACF}$ items. The model can then proceed by treating the access pattern as uniform, thus greatly simplifying the derivation.

For $b - c$ access, $ACF = \frac{\alpha}{D} = \left(1 + \frac{(b-c)^2}{c(1-c)}\right)/D$, so it is an aggregated parameter that combines parameters $b$, $c$ and $D$. This is a *bottom-up approach* to parameterizing the model.

To give another example, suppose a fraction $b$ of requests are writes, and the rest are reads. Under certain assumptions, one can show that the probability of a conflict over $D$ items is the same as that for write-only requests over $D/\beta$ items, where $\beta = 1 - (1-b)^2$ [56].

A bottom-up approach is not always feasible. For example, it is hard to see how we can derive a formula for ACF that expresses it in terms of parameters for the TPC-C workload in the TAS experiments. An alternative, *top-down approach* would be to *empirically* determine how the ACF depends on other model parameters (e.g., the number of records in the TPC-C tables), and thus refine it by expressing it in terms of those other parameters.

This top-down approach has been demonstrated in the case of a Cache Miss Equation: For applications that use garbage-collected languages (e.g., Java), the equation's initial parameters are later expressed in terms of the heap size [60]; for a cache that is shared by $K$ virtual machines, the refinement is in terms of $K$ [64].

*Fig. 4* confirms that ACF remains the same when the number of nodes increases, and despite a drastic change in application workload (TPC-C and Radargun) and system architecture (private cluster and EC2). This is an excellent example of an analytic validation, since the right-hand side of *Eq. (2)* can be empirically measured. It is also an example of a *scientific* result that is robust with respect to a change in *technology*.

# CHAPTER 11

# Exercises

The following are some exercises on the models discussed in this book.

1. [*StreamJoins*] J. Kang, J.F. Naughton and S.D. Viglas. Evaluating window joins over unbounded streams. In Proc. *ICDE*, 341–352 (2003).

   The tuples in a stream may have varying inter-arrival times, and the joining time may also vary from one tuple to another. This makes it necessary to have a buffer for storing arriving tuples as they wait their turn to be processed.

   Suppose we model the inter-arrival time and processing time of each tuple as exponentially distributed (with rates $\lambda_a$, $\lambda_b$, and $\mu$). Assuming there are no CPU and memory constraints, estimate how much memory space is occupied when joining two streams with window sizes $A$ and $B$.

2. [*MediaStreaming*] Y.-C. Tu, J. Sun, and S. Prabhakar. Performance analysis of a hybrid media streaming system. In *Proc. SPIE/ACM Multimedia Computing and Networking*, 69–82 (2004).

   (I) *Fig. 2* indicates a time $k_1$ (beyond $k_0$) where bandwidth usage of the CDN servers becomes 0. Give an equation for the part of the curve between $k_0$ and $k_1$.

   (II) A requesting peer may, after streaming has started, decide to abort the download (because the streaming quality is poor, say). Explain how you can modify *Eq. (1)* to model stream abortion.

   (III) What is the gradient (expressed in terms of model parameters) of the asymptotic line in *Fig. 3b* ("Total Peers" vs. "Time")?

3. [*WebCrawler*] J. Cho and H. Garcia-Molina. The evolution of the web and implications for an incremental crawler. In *Proc. Int. Conf. on Very Large Data Bases*, 200–209 (2000).

   Assume inter-update times are exponentially distributed.

   (I) Show that expected freshness $EF(c_k; t)$ decays exponentially.

   (II) Use the periodicity of freshness to show that $\frac{1}{I} \int_s^{s+I} EF(c_k; t) dt$ is the same for any $s$ (so batch and incremental crawling have the same average freshness).

   (III) Derive the equation for the line in *Fig. 6*.

(IV) Suppose the lifespan of a page is exponentially distributed with rate $\alpha$. If you have the data for figure below, how can you use that data to estimate $\alpha$?

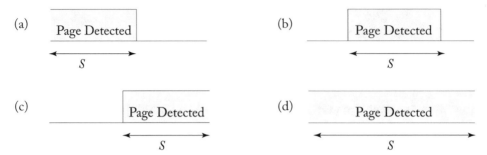

Figure 3: Issues in estimating the lifespan of a page

(V) If a web crawler visits a page before it is updated, that is wasted effort. On the other hand, if the visit happens after multiple updates, then the crawler has missed one or more updates. Assume inter-update time is exponentially distributed with rate $\lambda$. Suppose the time between two crawls to a given page is a constant $T$. Determine the probability that there is exactly one update between two crawls to a page. What is the value of $T$ that maximizes this probability?

4. [*SoftErrors*] A.A. Nair, S. Eyerman, L. Eeckhout and L.K. John. A first-order mechanistic model for architectural vulnerability factor. In Proc. *IEEE ISCA*, 273–284 (2012).

(I) Consider *Fig. 2(a)* for ROB occupancy in the shadow of a L2 miss. Let the increase in occupancy start at time $t_1$ and end at time $t_2$, as shown below.

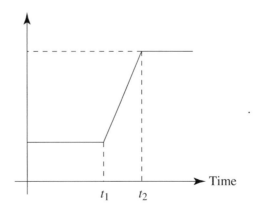

Before $t_1$, suppose an instruction stays in the buffer for time $R_1$ (on average); similarly, after $t_2$, suppose an instruction stays in the buffer for time $R_2$ (on average). Show that $R_2 - R_1 = t_2 - t_1$.

(II) Consider a microarchitecture with dispatch width $D$ and reorder buffer size $W$. Let the average instruction latency be $\ell$ cycles, and let $K(W)$ be the average critical path length of the instructions in the reorder buffer for a given program. Explain why execution is expected to stall if $D > \frac{W}{\ell K(W)}$.

(III) Consider the occupancy model in *Fig. 2(c)*. Explain why an instruction cache miss can lead to an empty reorder buffer if the $L2$ hit latency exceeds $W/D$.

5. [*SleepingDisks*] Q. Zhu, Z. Chen, L. Tan, Y. Zhou, K. Keeton, and J. Wikes. Hibernator: helping disk arrays sleep through the winter. In *Proc. SOSP*, 177–190 (2005).

In *Sec. 3.2.4*, the authors' expression for $R_{ij}'''$ uses $\frac{Exp(t_{ij})}{2}$ to estimate the delay caused by servicing one background request.

(I) Explain why this estimate is inconsistent with their $M/G/1$ model for the disk.

(II) Provide an alternative estimate.

6. [*StorageAvailability*] Gabber, Fellin, Flaster, Gu, Hillyer, Ng, Ozden, and Shriver. StarFish: highly-available block storage. In *Proc. USENIX Annual Technical Conference*, 151–163 (2003).

Suppose $N = 3$, the 3 SEs have different failure and recovery rates, and $Q = 1$. What is the availability of the system?

7. [*DatacenterAMP*] V. Gupta and R. Nathuji. Analyzing performance asymmetric multicore processors for latency sensitive datacenter applications. In *Proc. HotPower* (2010).

(I) Assume the fraction of computation that can be parallelized is $f = 0.8$, and $perf(r) = \sqrt{r}$. Let the processing speed for $r = 1$ be $\mu_1$. For SMP, how should an area of size $n$ be symmetrically divided to maximize the processing speed? Suppose the service level agreement specifies both latency $T_{SLA}$ and arrival rate $\lambda_{SLA}$. How big must $n$ be for SMP to satisfy this requirement?

(II) In practice, the SLA is specified not as a mean $T_{SLA}$, but with a percentile. For example, the specification has two values $T_{bound}$ and $\lambda_{SLA}$, and 98% of all requests must have response time smaller than $T_{bound}$ whenever arrival rate is smaller than $\lambda_{SLA}$. For an $M/M/1$ queue, it is known that the system time is exponentially distributed with mean $1/(\mu - \lambda)$ (cf. Eq. (2.6)); how can this fact be used to determine what $\lambda_{SLA}$ should be?

8. [*GPU*] J.-C. Huang, J.H. Lee, H. Kim, and H.-H. S. Lee. GPUMech: GPU performance modeling technique based on interval analysis. In *Proc. IEEE/ACM MICRO*, 268–279 (2014).

In *Sec. IV-B1*, the authors model queueing delay for MSHR entries using *Eq. (19)*. Consider the following alternative queueing models.

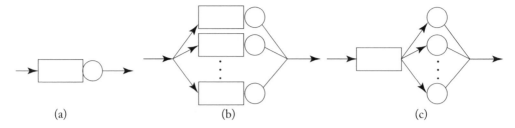

(a)                                    (b)                                    (c)

(I) Explain why (a) and (b) are not suitable alternatives to *Eq. (19)*.

(II) Suppose we use (c) and $M/M/k$ as the queueing model. What is the value for $k$, and how can we estimate arrival and service rates?

9. [*GPRS*] G. Nogueira, B. Baynat, and P. Eisenmann. An analytical model for the dimensioning of a GPRS/Edge network with a capacity constraint on a group of cells. In *Proc. MobiCom*, 215–227 (2005). In their ON/OFF model, an ON period corresponds to service time and an OFF period corresponds to inter-arrival time.

(I) Using the variables in their GPRS model and slots as the unit of time (instead of seconds), estimate the average inter-arrival time and service time of transfers.

(II) Deduce the arrival rate and service rate of transfers.

(III) Give an interpretation for $x$ in *Eq. (6)*.

10. [*TransactionalMemory*] A. Heindl, G. Pokam, and A.-R. Adl-Tabatabai. An analytic model of optimistic software transactional memory. In *Proc. IEEE Int. Symp. on Performance Analysis of Systems and Software (ISPASS)*, 153–162 (2009).

Suppose we remove the absorbing state $k + 1$ in *Fig. 1* by adding a transition, with probability 1, from state $k + 1$ to state 0.

(I) Write steady state balance equations for this equivalent model.

(II) Show that
$$p_i = p_0 q_0 q_1 \cdots q_{i-1} \quad \text{for } i = 1, 2, \ldots, k + 1.$$

(III) Deduce *Eq. (8)* from (II).

11. [*RouterBuffer*] G. Appenzeller, I. Keslassy and N. McKeown, Sizing router buffers, In *Proc. SIGCOMM*, 281-292 (2004).

(I) Consider the following argument for sizing the buffer:

"When the bottleneck link is fully utilized, the throughput is the link's data rate $C$. By Little's Law, the amount of data in the network is $C \times \overline{RTT}$, where $\overline{RTT}$ is the average round-trip time. To avoid packet loss, the bottleneck link should have buffer size $B = C \times \overline{RTT}$."

What is wrong with this argument?

(II) For short flows, the authors estimate queue length by

$$E[Q] = \frac{\rho}{2(1-\rho)} \frac{E[X^2]}{E[X]} + \epsilon, \quad \text{where } \epsilon = -\rho \frac{E[X]}{2}.$$

Explain the factor $\rho$ in $\epsilon$. Why does this correction term $\epsilon$ implicitly assume the queue is $M/D/1$.

(III) In TCP's congestion avoidance phase, the congestion window size $W$ is controlled by additive increase and multiplicative decrease (AIMD). In particular, if there is a congestion signal (e.g., missing acknowledgement), the window size will be reduced from $W$ to $\alpha W$, where $0 < \alpha < 1$. For a single long-lived TCP flow, the paper shows that buffer size should be $B = \overline{RTT} \times C$ if $\alpha = \frac{1}{2}$. What should $B$ be if $\alpha = \frac{1}{3}$?

12. [*ProactiveReplication*] A. Duminuco, E. Biersack, and T. En-Najjary. Proactive replication in distributed storage systems using machine availability estimation. In *Proc. CoNEXT*, 27:1–27:12 (2007).

The following is a clearer version of *Fig. 3*:

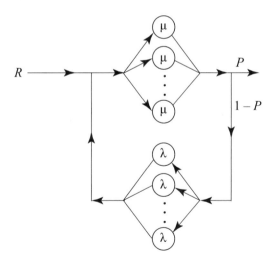

(I) Assume the inter-repair time and the time a peer stays connected or (temporarily) disconnected are exponentially distributed. Let $p_{n,m}$ be the probability that there are $n$ peers in the *connected* state and $m$ peers in the *disconnected* state. Show how you can use local balance to derive

(a) $p_{n,m}(1 - P)n\mu = p_{n-1,m+1}(m + 1)\lambda$;

(b) $p_{n,m}R = p_{n+1,m}P(n + 1)\mu$.

(II) Draw a Markov chain for the state $\langle n, m \rangle$. Give an example of a flow balance equation for your Markov chain.

13. [*InternetServices*] B. Urgaonkar, G. Pacific, P. Shenoy, M. Spreitzer, and A. Tantawi. An analytical model for multi-tier Internet services and its applications. In *Proc. SIGMET-RICS*, 291–302 (2005).

    Consider the experiment for *Fig. 9*. For the case of 400 simultaneous sessions, estimate:

    (I) the service demand at the middle tier;

    (II) the goodput at the CPU in the middle tier;

    (III) the average number of sessions that are waiting for response to their requests;

    (IV) the arrival rate of requests; and

    (V) the probability that a request is dropped (i.e., not completed).

14. [*SensorNet*] R.C. Shah, S. Roy, S. Jain, and W. Brunette. Data mules: Modeling a three-tier architecture for sparse sensor networks. In *Proc. IEEE Workshop on Sensor Network Protocols and Applications*, 30–41 (2003).

    Suppose there are no access points. Instead, MULEs arrive at rate $\lambda$ from outside the sensor network and move about in a random walk to collect data. After an exponentially distributed time $T$, a MULE stops its collection and leaves the network.

    (I) Draw a Markov chain for the number of MULEs in the network.

    (II) What is the probability that there are no MULEs in the network?

    (III) Explain why the expected number of data units collected by a MULE does not depend on $ET$.

15. [*DependabilitySecurity*] K.S. Trivedi, D.S. Kim, A. Roy, and D. Medhi. Dependability and security models. In *Proc. Design of Reliable Communication Networks*, 11–20 (2009).

    (I) Consider the Workstation File Server System in *Fig. 12* below:

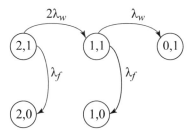

Figure 12: PR model of Workstation File server system

There is only one system administrator who can repair the file server and workstations, and repairing the file server has priority over repairing the workstations. The average repair time is $T_f$ for the file server and $T_w$ for a workstation. Draw the Markov chain for this system and write down enough equations for solving this system.

(II) Consider a network of 7 sensors for *Fig. 13*. A sensor is secure and cannot be captured unless it has failed. The failure rate is $\lambda$, and failures cannot be repaired. A failed sensor can be captured at rate $\gamma$, and captured sensors cannot recover. If more that three sensors are captured, the sensor network itself is considered to have failed, i.e., unusable. Draw a Markov chain to model such a system.

16. [*ServerEnergy*] B. Guenter, N. Jain and C. Williams. Managing cost, performance, and re-liability tradeoffs for energy-aware server provisioning. In *Proc. IEEE INFOCOM*, 1332–1340 (2011).

Consider the Markov state diagram in *Fig. 6(a)*.

(I) Show that, in steady state, the (discretized) time a server is "on" has a Geometric ($p$) distribution. Suggest a way of estimating $p$.

(II) *Fig. 6(a)* is an unrolled version of the state diagram in *Fig. 4(a)*. Draw the unrolled version of the state diagram for *Fig. 4(b)*, for time steps $1, 2, \ldots, 13$. Let $s_i, h_i, o_i$, and $f_i$ denote the *suspend*, *hibernate*, *on* and *off* power states in time step $i$. Write the balance equations for the number of servers in each state for time Step 8.

17. [*OpenClosed*] B. Schroeder, A. Wierman and M. Harchol-Balter. Open vs. closed: a cau-tionary tale. In *Proc. NSDI* (2006).

The paper implicitly assumes the probability $p$ (that a user makes another request after a request completes) is constant.

Suppose $p$ decreases as load increases.

(I) Explain why this may happen.

(II) Suggest how **Principle (vii)** should be modified:

    **Principle (vii)'**   As load increases, …

18. [*NetworkProcessor*] J. Lu and J. Wang. Analytical performance analysis of network-processor-based application designs. In *Proc. Int. Conf. Computer Communications and Net-works*, 33–39 (2006).

With bounded queue length, packets are dropped when the input buffer is full. This may cause the packet source to retransmit the packet. The arrival rate $\lambda$ should therefore include retransmitted packets, so $\lambda$ depends on the packet rejection probability $p_K$ in *Eq. (3-17)*.

(I) How would you model the relationship between $\lambda$ and $p_K$?

(II) Suppose the input buffer is finite and dropped packets are retransmitted. Suggest how the network processor model can be solved, using a fixed-point approximation or some other method.

19. [*PipelineParallelism*] A. Navarro, R. Asenjo, S. Tabik, and C. Cascaval. Analytical modeling of pipeline parallelism, In *Proc. PACT*, 281–290 (2009).

(I) Explain *Eq. (3)*.

(II) Consider a special case of *Fig. 5(a)*, where there are just two queues processing 6 items: one queue has two servers, the other queue has three servers, and service times are exponentially distributed with rates $\mu_A$ and $\mu_B$, as indicated below:

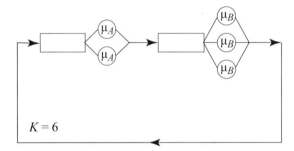

Draw the state transition diagram for this model and write down enough equations for solving this birth-death process.

(III) The collapsed stage model for `ferret` in *Sec. 3.1* is as follows:

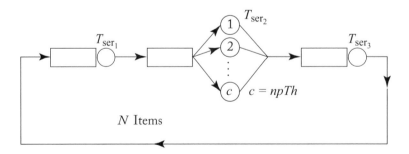

The input stage has average service time $T_{ser_1}$, the collapsed stage has $c = npTh$ servers with average service time $T_{ser_2}$ each, the output stage has average service time $T_{ser_3}$, and the number of items is $N(\geq c)$. Instead of using *Eq. (3)*, expain how you can solve this model approximately with hierarchical decomposition.

(IV) Show how the hierarchical decomposition for *Fig. 5(a)* can be done differently, using iterated decomposition.

20. [*DatabaseScalability*] S. Elnikety, S. Dropsho, E. Cecchet, and W. Zwaenepoel. Predicting replicated database scalability from standalone database profiling. In *Proc. ACM EuroSys*, 303–316 (2009).

    (I) Explain the statement in *Sec. 3.3.1*: "To successfully commit $W$ update transactions, $W/(1 - A_1)$ update transactions are submitted."

    (II) Suppose the number of replicas ($N$) is fixed, but the number of clients per replica ($C$) is increased. How would that affect the abort probability $A_N$? Explain.

    (III) For the multimaster model in *Fig. 1*, let $B$ denote the average load balancer and network delay and let $C$ denote the certification delay (per attempt). For an $N$-replica system, suppose $R_N$ is the average response time obtained by MVA for a single replica (containing one CPU and one disk). What is the average transaction response time?

    (IV) For a multi-master replicated database, suppose each replica in *Fig. 1* has two disks (instead of one). How does that affect the MVA calculation?

    (V) The conflict window $CW(N)$ cannot be measured on a stand-alone system. Suppose Schweitzer's approximation is used in the MVA algorithm. Suggest how $CW(N)$ can be estimated by a fixed-point approximation.

21. [*TCP*] J. Padhye, V. Firoiu, D. Towsley, and J. Kurose. Modeling TCP throughput: a simple model and its empirical validation. In *Proc. SIGCOMM*, 303–314 (1998).

    Give an AVA (Average Value Approximation) argument for *Eq. (21)*:

    $$B = \frac{E[Y] + Q * E[R]}{E[A] + Q * E[Z^{TO}]}.$$

    (You can ignore the subscript $i$.)

22. [*DatabaseSerializability*] P.A. Bernstein, A. Fekete, H. Guo, R. Ramakrishnan, and P. Tamma. Relaxed currency serializability for middle-tier caching and replication. In *Proc. SIGMOD*, 599–610 (2006).

    Suppose the equilibrium is decomposed into demand $\lambda_{in}$ and supply $\lambda_{out}$ (see Chapter 8).

    (I) How does the currency bound affect supply $\lambda_{out}$?

    (II) Assume a flash crowd is modeled by an increase in num_terminals (*Table 1*). How does that affect the demand line $\lambda_{in}$?

    (III) The idea in RC-Serializability is to trade currency for performance—a more relaxed currency constraint will help improve performance. Intuitively, a more relaxed currency constraint can make performance less sensitive to flash crowds. Use the decomposition into $\lambda_{in}$ and $\lambda_{out}$ to explain this intuition.

23. [*Roofline*] S. Williams, A. Waterman, and D. Patterson. Roofline: an insightful visual performance model for multicore architectures. *CACM 52(4)*, 65–76 (2009).

    (I) Explain why the sloping rooflines are parallel.

    (II) Suppose $\mathcal{P}$ and $\mathcal{Q}$ are two processors from the same architecture family (e.g., same instruction set). $\mathcal{P}$ has higher peak memory bandwidth than $\mathcal{Q}$, but a slower clock. Explain how you can use the roofline model to choose between $\mathcal{P}$ and $\mathcal{Q}$ when given a particular kernel.

    (III) Discuss how the roofline and operational intensity may (or may not) be affected in the following cases: (a) An L3 cache is added. (b) An asymmetric multicore processor is reconfigured (as described in *DatacenterAMP* [22]) to reduce the size of its large core.

24. [*PerformanceAssurance*] N. Roy, A. Dubey, A. Gokhale, and L. Dowdy. A capacity planning process for performance assurance of component-based distributed systems. In *Proc. ACM/SPEC ICPE*, 259–270 (2011).

    (I) Consider a separable closed queueing network model like the one shown in *Fig. 2*. Let $N$ be the number of clients. It is known that the throughout $X(N)$ for such a network increases as $N$ increases. Explain why this implies that the utilization for any queue also increases as $N$ increases.

    (II) The paper replaces service demand $D$ by *overall service demand OSD*, to model the increase in overhead when utilization increases (like *Eq. (2)* and *Eq. (3)*). Explain how, using MVA, this can cause the calculated throughput to decrease when the number of clients increases.

    (III) In *Sec. 2.2.2*, the paper approximates a queue that has $k$ servers, each with service rate $\mu$, by a load-dependent queue with 1 server and service rate $f_{\text{CI}}\mu$, where $f_{\text{CI}}$ is *CI*, the inverse of the correction factor. Explain why, in *Fig. 9*, $f_{\text{CI}}$ converges to $k$ (the number of processors) as the number of clients increases.

25. [*BitTorrent*] D. Qiu and R. Srikant. Modeling and performance analysis of BitTorrent-like peer-to-peer networks. In *Proc. SIGCOMM*, 367–378 (2004).

    Let $T_{\text{aborted}}$ and $T_{\text{completed}}$ be the average downloading time for a download that is aborted or completed, respectively. Show that:

    (I) $\bar{x} = (\lambda - \theta\bar{x})T_{\text{completed}} + \theta\bar{x}T_{\text{aborted}}$;

    (II) the fraction of downloads that will become seeds is $\frac{\lambda - \theta\bar{x}}{\lambda}$ if and only if $T_{\text{completed}} = T_{\text{aborted}}$.

26. [*NoC*] J. Kim, D. Park, C. Nicopoulos, N. Vijaykrishnan, and C.R. Das. Design and analysis of an NoC architecture from performance, reliability and energy perspective. In *Proc. Symp. Architecture for Networking and Commun. Systems*, 173–182 (2005).

    Give an example of Average Value Approximation (AVA) in this paper.

27. [*Gossip*] R. Bakhshi, D. Gavidia, W. Fokkink, and M. van Steen. An analytical model of information dissemination for a gossip-based protocol. In *Proc. Distributed Computing and Networking*, 230–242 (2009).

(I) Explain why the equation (in *Sec. 3.3*)

$$E[P_{drop}] = \frac{s-k}{s-\hat{s}} = \frac{s - s \cdot \frac{c}{n}}{s - s \cdot \frac{c}{n} \cdot \frac{s}{c}} = \frac{n-c}{n-s}$$

is an example of Average Value Approximation (AVA).

(II) *Eq. (3)* has three parameters: $s$, $c$, and $n$. Show how this equation can be rewritten, to give an expression for $\frac{dx}{dt}$ that has just two parameters.

(III) The analysis in *Sec. 3.4* says convergence is fastest when $s = n - \sqrt{n(n-c)}$, and *Sec. 4* verifies this for just one case: $c = 100$ and $n = 500$. Suggest how you can verify the claim for a larger variety of values for $c$ and $n$.

(IV) Explain how you can analytically validate *Eq. (4)*:

$$x(t) = \frac{e^{\alpha t}}{(N - \frac{n}{c}) + \frac{n}{c} e^{\alpha t}}.$$

28. [*CachingSystems*] V. Martina, M. Garetto, and E. Leonardi. A unified approach to the performance analysis of caching systems. In *Proc. IEEE INFOCOM*, 2040–2048 (2014).

(I) The authors used renewal theory to derive some of the expressions in the paper. Give an example of an expression that can be derived with AVA (Average Value Approximation), instead of using renewal theory.

(II) Explain why PASTA implies $p_{in}(m) = p_{hit}(m)$.

(III) Explain how you can analytically validate (see Sec. 9.2.8) the small cache LRU approximation (in *Sec. IV-G*)

$$p_{hit}(m) \approx (\lambda_m T_C) - \frac{(\lambda_m T_C)^2}{2}.$$

29. [*CodeRed*] C.C. Zou, W. Gong, and D. Towsley. Code Red worm propagation modeling and analysis. In *Proc. CCS*, 138–147 (2002).

In the worm model, one factor that reduces the infection rate $\beta$ is network congestion caused by worm propagation.

(I) Suggest how an $M/M/1$ queue can be used to model congestion delay.

(II) Express $\beta$ in terms of the queue's arrival and service rates.

(III) Explain why your $M/M/1$ queue may be a poor model for congestion delay.

30. [*DistributedProtocols*] I. Gupta. On the design of distributed protocols from differential equations. In *Proc. ACM Symp. Principles of Distributed Computing*, 216–225 (2004).

(I) The state machine in *Fig. 1* can be viewed as a Markov chain. For this Markov chain, what do $x$, $y$, and $z$ mean and what are the transition rates?

(II) What is the average length of each line in *Fig. 8*?

(III) Explain why *Eq. (7)* is equivalent to

$$\begin{aligned}
\dot{x} &= 0.3xz - 0.3xy \\
\dot{y} &= 0.3yz - 0.3xy \\
\dot{z} &= -0.3xz - 0.3yz + 0.3xy + 0.3xy.
\end{aligned}$$

(The advantage here is that 0.3 can be interpreted as a probability for flipping and one-time-sampling.)

31. [*EpidemicRouting*] X. Zhang, G. Neglia, J. Kurose, and D. Towsley. Performance modeling of epidemic routing. In *Proc. Networking*, 2867–2891 (2007).

(I) For the routing scheme in *Sec. 2.2*, estimate the number of hops taken by a delivered packet.

(II) Consider the 2-hop forwarding scheme in *Sec. 3*. Suppose, to improve the delivery rate, the protocol induces an infected node to (somehow) enlist more nodes to enter the system, and we model this as

$$N'(t) = \gamma I(t) \qquad \text{for some positive real value } \gamma.$$

Use eigenvalues to examine how the system behaves as $t \to \infty$.

(III) Let $N$ be the number of nodes (excluding the destination), and $I(t)$ the number of "infected" nodes carrying a copy of a packet. Then 2-hop forwarding can be modeled as

$$\frac{dI(t)}{dt} = \beta S(t) \quad \text{where } S(t) = N - I(t) \text{ and } I(0) = 1.$$

Formulate this model as an ordinary differential equation for the vector $\mathbf{u}(t) = \begin{pmatrix} S(t) \\ I(t) \end{pmatrix}$.

Determine the eigenvalues and describe how $\mathbf{u}(t)$ moves in the $S$-$I$ plane over time $t$.

32. [*InformationDiffusion*] Y. Matsubara, Y. Sakurai, B.A. Prakash, L. Li, and C. Faloutsos. Rise and fall patterns of information diffusion: model and implications. In *Proc. ACM KDD*, 6–14 (2012).

(I) Show that *Eq. (7)* has a special case that is the discretized form of *Eq. (1)*.

(II) For (I), show that $\Delta B(n)$ increases to a maximum around $B(n) = \frac{N}{2}$, then starts to decrease after that.

33. [*WirelessCapacity*] P. Gupta and P.R. Kumar. The capacity of wireless networks. *IEEE Trans. on Information Theory*, 388–404 (2000).

    Assume the domain has area $A$ m$^2$, instead of 1 m$^2$.

    (I) For the Protocol Model, point out where $A$ enters the derivation.

    The paper assumes $A$ is fixed when the number of nodes $n$ is increased, so density (#nodes per square meter) increases.

    (II) Suppose density is fixed when $n$ increases, so $A$ increases. What would be the new bound for $\lambda \bar{L}$?

    (III) Explain intuitively why this new bound is $O(1)$ instead of $O(\frac{1}{\sqrt{n}})$.

34. [*NonstationaryMix*] C. Stewart, T. Kelly, and A. Zhang. Exploiting nonstationarity for performance prediction. In *Proc. EuroSys*, 31–44 (2007).

    By considering the number of visits per type $j$ transaction to resource $r$, and the service time at resource $r$, justify the term

    $$\sum_r \left( \frac{1}{\lambda_i} \frac{U_{ir}^2}{1 - U_{ir}} \right) \sum_{j=1}^{n} N_{ij}$$

    in *Eq. (4)*.

35. [*P2PVoD*] B. Fan, D.G. Andersen, M. Kaminsky, and K. Papagiannaki. Balancing throughput, robustness, and in-order delivery in P2P VoD. In *Proc. CoNEXT* (2010).

    (I) Explain why the robustness definition $R = 1 - p^{\bar{r}}$ is an example of AVA (Average Value Approximation).

    (II) The analysis for the hybrid strategy leads to

    $$T^{hybrid} \approx \frac{S^{hybrid}}{(1 - S^{hybrid}) \ln \frac{1}{1 - S^{hybrid}}} \left( \frac{bM}{U_p} - \frac{U_s}{\lambda U_p} \right).$$

    Suggest an analytic validation of this approximation.

36. [*MapReduce*] E. Vianna, G. Comarela, T. Pontes, J. Almeida, V. Almeida, K. Wilkinson, H. Kuno, and U. Dayal. Analytical performance models for MapReduce workloads. *Int. J. Parallel Programming*, 41(4):495–525, (2013).

    (I) Explain why *Eq. (6)* can be rewritten as

    $$A_{ik}(\vec{N}) \approx \sum_{j=1}^{C} f_{ij} Q_{jk}(\overrightarrow{N - 1_i}),$$

for some $f_{ij}$, and how this modification to *Eq. (2)* can be interpreted.

(II) Consider a MapReduce job where a dataset is divided among $m$ map tasks. Explain how resource contention and precedence constraint can interact to increase or decrease average job response time when $m$ increases.

(III) Explain why task precedence (e.g., *Fig. 5*) can cause average resource utilization to flatten out before 100% for the 4-node experiment in *Fig. 8*.

37. [*802.11*] Y.C. Tay and K.C. Chua. A capacity analysis for the IEEE 802.11 MAC protocol. *Wireless Networks*, 159–171 (2001).

It follows from *Claim 5* that, for a fixed bandwidth and sufficiently large packet size $\ell$ (measured in #slots), the saturation throughput is maximum when $W \approx (n-1)\sqrt{\ell}$. Suggest how the simulator can be used to validate this.

38. [*SoftState*] J.C.S. Lui, V. Misra and D. Rubenstein. On the robustness of soft state protocols. In *Proc. IEEE Int. Conf. Network Protocols*, 50–60 (2004).

(I) Suggest how you can perform an analytic validation of *Eq. (1)* for a soft-state protocol $\mathcal{P}$. You may assume $a = b = 1$.

(II) In *Sec. 4*, the client for a hard-state protocol tries $n$ times to tell the server that it wants to terminate the connection. Explain why *Eq. (4)* may not be a good model for the server's failure to receive the notification.

39. [*CloudTransactions*] D. Kossman, T. Kraska, and S. Loesing. An evaluation of alternative architectures for transaction processing in the cloud. In *Proc. SIGMOD*, 579–590 (2010).

(I) Consider *Fig. 7*, which plots WIPS against EB. Note that the curves for all architectures share the same gradient for small EB. Deduce from this that the database server in the experiments is a commodity machine (i.e., not faster).

(II) A TPC-W request is considered *valid* (and counted in WIPS) only if its response time is below a specified threshold. Sketch how the WIPS-vs.-EB curve for RDS is affected if the threshold is lowered.

(III) An EB simulates a user who issues a request, waits for an answer, then issues the next request after a think time (about 7 sec). Thus, if there are $N$ EBs, then there are $K$ users waiting for an answer and $N - K$ users who are thinking. Given a WIPS-vs.-EB curve, $N$, and average think time $Z$, explain how you can determine $K$.

(IV) Consider the throughput (WIPS in RT) for Google AppEngine. On *Fig. 10*, sketch the new throughput curve if the Master database server is replaced by a machine that is more powerful.

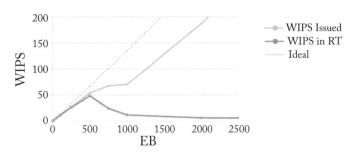

Figure 10: AE/C: WIPS(EB)

(V) Suppose the system is in a stable low equilibrium at the tail of the throughput curve in *Fig. 10*. Explain how increasing the power of the Master can shift the system to a stable high equilibrium.

(VI) Use an open model to estimate the transaction response time for the partition architecture (in terms of transaction arrival rate, number of servers, service times per visit to a server and number of visits per transaction to a server).

40. [*ElasticScaling*] D. Didona, P. Romano, S. Peluso, and F. Quaglia. Transactional Auto Scaler: Elastic scaling of in-memory transactional data grids. In *Proc. ICAC*, 125–134 (2012).

(I) Describe a set of experiments and measurements for generating *Fig. 4*.

(II) Let $X$ be the random variable for a requested item, $\lambda(X)$ the arrival rate of lock requests for $X$, and $T(X)$ the average (over all requests for $X$) lock holding time for $X$. Describe how you would measure $\lambda(X)$ and $T(X)$. Let $\lambda = E\lambda(X)$, $T = ET(X)$ and $U$ be the mean lock utilization per item. Explain why $U = \lambda T$ is an average value approximation (AVA) when the lock requests are not uniformly distributed over the data items.

# Bibliography

[1] G. Appenzeller, I. Keslassy, and N. McKeown. Sizing router buffers. *SIGCOMM Comput. Commun. Rev.*, 34(4):281–292, August 2004. DOI: 10.1145/1030194.1015499. 30, 33, 57, 68, 94, 112, 153

[2] R. Bakhshi, D. Gavidia, W. Fokkink, and M. Steen. An analytical model of information dissemination for a gossip-based protocol. In *Proc. Int. Conf. Distributed Computing and Networking (ICDCN)*, pages 230–242, Berlin, Heidelberg, 2009. Springer-Verlag. DOI: 10.1007/978-3-540-92295-7_29. 67, 71, 79, 81, 96, 113, 151

[3] P. A. Bernstein, A. Fekete, H. Guo, R. Ramakrishnan, and P. Tamma. Relaxed-currency serializability for middle-tier caching and replication. In *Proc. ACM SIGMOD Int. Conf. Management of Data*, pages 599–610, June 2006. DOI: 10.1145/1142473.1142540. 56, 58, 59, 77, 93, 110, 150

[4] T. Camp, J. Boleng, and V. Davies. A survey of mobility models for ad hoc network research. *Wireless Communications & Mobile Computing (WCMC): Special Issue on Mobile Ad Hoc Networking: Research, Trends and Applications*, 2:483–502, 2002. DOI: 10.1002/wcm.72. 93, 151

[5] N. Cardwell, Y. Cheng, C. S. Gunn, S. H. Yeganeh, and V. Jacobson. BBR: congestion-based congestion control. *Commun. ACM*, 60(2):58–66, 2017. DOI: 10.1145/3009824. 53

[6] A. Chesnais, E. Gelenbe, and I. Mitrani. On the modeling of parallel access to shared data. *Commun. ACM*, 26(3):196–202, March 1983. DOI: 10.1145/358061.358073. 104, 150, 151

[7] J. Cho. *Crawling the Web: Discovery and Maintenance of a Large-Scale Web Data*. Ph.D. thesis, Stanford University, November 2001. 12

[8] J. Cho and H. Garcia-Molina. The evolution of the web and implications for an incremental crawler. In *Proc. Int. Conf. on Very Large Data Bases (VLDB)*, pages 200–209, San Francisco, CA, 2000. Morgan Kaufmann Publishers, Inc. 12, 94, 111, 153

[9] P. J. Courtois. Decomposability, instabilities, and saturation in multiprogramming systems. *Commun. ACM*, 18(7):371–377, 1975. DOI: 10.1145/360881.360887. 55, 74, 150

[10] P. J. Denning and J. P. Buzen. The operational analysis of queueing network models. *ACM Comput. Surv.*, 10(3):225–261, 1978. DOI: 10.1145/356733.356735. 45, 100, 150

[11] D. Didona, P. Romano, S. Peluso, and F. Quaglia. Transactional auto scaler: elastic scaling of in-memory transacti onal data grids. In *Proc. Int. Conf. Autonomic Computing (ICAC)*, pages 125–134, September 2012. DOI: 10.1145/2371536.2371559. 108, 115, 120, 150

[12] L. Dowdy and C. Lowery. *P.S. to Operating Systems*. Prentice-Hall, Inc., Upper Saddle River, NJ, 1993. 39, 151

[13] A. Duminuco, E. Biersack, and T. En-Najjary. Proactive replication in distributed storage systems using machine availability estimation. In *Proc. ACM CoNEXT Conf.*, pages 27:1–27:12, December 2007. DOI: 10.1145/1364654.1364689. 30, 33, 42, 79, 152

[14] S. Elnikety, S. Dropsho, E. Cecchet, and W. Zwaenepoel. Predicting replicated database scalability from standalone database profiling. In *Proc. ACM EuroSys Conf.*, pages 303–316, April 2009. DOI: 10.1145/1519065.1519098. 46, 48, 57, 96, 105, 115, 150

[15] S. Eyerman, L. Eeckhout, T. Karkhanis, and J. E. Smith. A mechanistic performance model for superscalar out-of-order processors. *ACM Trans. Comput. Syst.*, 27(2):3:1–3:37, May 2009. DOI: 10.1145/1534909.1534910. 79

[16] B. Fan, D. G. Andersen, M. Kaminsky, and K. Papagiannaki. Balancing throughput, robustness, and in-order delivery in P2P VoD. In *Proc. Co-NEXT*, pages 10:1–10:12, New York, 2010. ACM. DOI: 10.1145/1921168.1921182. 90, 100, 113, 152

[17] S. Floyd, M. Handley, J. Padhye, and J. Widmer. TCP Friendly Rate Control (TFRC): Protocol Specification. RFC 5348, September 2008. DOI: 10.17487/rfc5348. 109

[18] E. Gabber, J. Fellin, M. Flaster, F. Gu, B. Hillyer, W. T. Ng, B. Özden, and E. A. M. Shriver. Starfish: highly-available block storage. In *Proc. USENIX Annual Tech. Conf.*, pages 151–163, June 2003. 20, 21, 31, 56, 68, 92, 153

[19] B. Guenter, N. Jain, and C. Williams. Managing cost, performance, and reliability trade-offs for energy-aware server provisioning. In *Proc. IEEE INFOCOM*, pages 1332–1340, April 2011. DOI: 10.1109/INFCOM.2011.5934917. 40, 41, 69, 96, 153

[20] I. Gupta. On the design of distributed protocols from differential equations. In *Proc. ACM Symp. on Principles of Distributed Computing (PODC)*, pages 216–225, July 2004. DOI: 10.1145/1011767.1011799. 77, 80, 111, 150

[21] P. Gupta and P. R. Kumar. The capacity of wireless networks. *IEEE Trans. Informa. Theory*, 46(2):388–404, 2000. DOI: 10.1109/18.825799. 7, 90, 98, 153

[22] V. Gupta and R. Nathuji. Analyzing performance asymmetric multicore processors for latency sensitive datacenter applications. In *Proc. Int. Conf. on Power Aware Computing and Systems (HotPower)*, pages 1–8, Berkeley, CA, 2010. USENIX Association. 20, 21, 94, 112, 132, 150

[23] M. Harchol-Balter. *Performance Modeling and Design of Computer Systems: Queueing Theory in Action*. Cambridge University Press, Cambridge, UK, 2013. DOI: 10.1017/CBO9781139226424. 28, 44, 151

[24] A. Heindl, G. Pokam, and A.-R. Adl-Tabatabai. An analytic model of optimistic software transactional memory. In *Proc. IEEE Int. Symp. on Performance Analysis of Systems and Software (ISPASS)*, pages 153–162, April 2009. DOI: 10.1109/ISPASS.2009.4919647. 30, 32, 60, 68, 80, 92, 110, 153

[25] M. D. Hill and M. R. Marty. Amdahl's law in the multicore era. *IEEE Computer*, 41(7):33–38, 2008. DOI: 10.1109/MC.2008.209. 21, 22

[26] J. Huang, J. H. Lee, H. Kim, and H. S. Lee. GPUMech: GPU performance modeling technique based on interval analysis. In *Proc. IEEE/ACM Int. Symp. Microarchitecture (MICRO)*, pages 268–279, December 2014. DOI: 10.1109/MICRO.2014.59. 20, 22, 31, 96, 114, 151

[27] J. R. Jackson. Jobshop-like queueing systems. *Manag. Sci.*, 10(1):131–142, 1963. DOI: 10.1287/mnsc.10.1.131. 29, 151

[28] J. Kang, J. F. Naughton, and S. Viglas. Evaluating window joins over unbounded streams. In *Proc. Int. Conf. on Data Engineering (ICDE)*, pages 341–352, March 2003. DOI: 10.1109/ICDE.2003.1260804. 8, 11, 20, 22, 67, 90, 108, 153

[29] J. Kim, D. Park, C. Nicopoulos, N. Vijaykrishnan, and C. R. Das. Design and analysis of an NoC architecture from performance, reliability and energy perspective. In *Proc. ACM Symp. Architecture for Networking and Communications Systems (ANCS)*, pages 173–182, October 2005. DOI: 10.1145/1095890.1095915. 67, 70, 94, 110, 151

[30] L. Kleinrock. *Queueing Systems, Vol. 1*. John Wiley, New York, 1975. 25, 151

[31] D. Kossmann, T. Kraska, and S. Loesing. An evaluation of alternative architectures for transaction processing in the cloud. In *Proc. SIGMOD*, pages 579–590, New York, NY, 2010. ACM. DOI: 10.1145/1807167.1807231. 108, 118, 149

[32] S. S. Lavenberg. Closed multichain product form queueing networks with large population sizes. In R. L. Disney and T. J. Ott, editors, *Applied Probability-Computer Science: The Interface, Vol. 1*, pages 219–249. Birkhauser, 1982. DOI: 10.1007/978-1-4612-5791-2. 56, 151

[33] S. S. Lavenberg and M. Reiser. Stationary state probabilities at arrival instants for closed queuing networks with multiple types of customers. *J. Appl. Probab.*, 17:1048–1061, 1980. DOI: 10.2307/3213214. 44, 149

[34] J. Lu and J. Wang. Analytical performance analysis of network processor-based application design. In *Proc. Int. Conf. Computer Communications and Networks*, pages 33–39, October 2006. DOI: 10.1109/ICCCN.2006.286241. 46, 47, 56, 58, 93, 151

[35] J. C. S. Lui, V. Misra, and D. Rubenstein. On the robustness of soft state protocols. In *Proc. IEEE Int. Conf. Network Protocols (ICNP)*, pages 50–60, October 2004. DOI: 10.1109/ICNP.2004.1348084. 108, 117, 153

[36] V. Martina, M. Garetto, and E. Leonardi. A unified approach to the performance analysis of caching systems. In *Proc. IEEE INFOCOM*, pages 2040–2048, April 2014. DOI: 10.1109/INFOCOM.2014.6848145. 67, 71, 97, 115, 121, 149

[37] Y. Matsubara, Y. Sakurai, B. A. Prakash, L. Li, and C. Faloutsos. Rise and fall patterns of information diffusion: Model and implications. In *Proc. ACM KDD*, pages 6–14, August 2012. DOI: 10.1145/2339530.2339537. 77, 81, 97, 116, 151

[38] R. J. T. Morris and W. S. Wong. Performance analysis of locking and optimistic concurrency control algorithms. *Perform. Eval.*, 5(2):105–118, May 1985. DOI: http://dx.doi.org/10.1016/0166-5316(85)90043-4. 106, 150

[39] R. Muntz and J. Wong. Asymptotic properties of closed queueing network models. In *8th Annual Princeton Conf. on Information Sciences and Systems*, March 1974. 53, 149

[40] A. A. Nair, S. Eyerman, L. Eeckhout, and L. K. John. A first-order mechanistic model for architectural vulnerability factor. In *Proc. IEEE Int. Symp. Computer Architecture (ISCA)*, pages 273–284, June 2012. DOI: 10.1109/ISCA.2012.6237024. 13, 23, 69, 79, 96, 114, 115, 153

[41] A. Navarro, R. Asenjo, S. Tabik, and C. Cascaval. Analytical modeling of pipeline parallelism. In *Proc. Int. Conf. on Parallel Architectures and Compilation Techniques (PACT)*, pages 281–290, September 2009. DOI: 10.1109/PACT.2009.28. 46, 48, 57, 78, 95, 112, 152

[42] G. Nogueira, B. Baynat, and P. Eisenmann. An analytical model for the dimensioning of a GPRS/EDGE network with a capacity constraint on a group of cells. In *Proc. MOBICOM*, pages 215–227, August 2005. DOI: 10.1145/1080829.1080852. 7, 30, 31, 39, 40, 41, 68, 91, 104, 109, 151

[43] J. Padhye, V. Firoiu, D. Towsley, and J. Kurose. Modeling TCP throughput: a simple model and its empirical validation. In *Proc. SIGCOMM*, pages 303–314, September 1998. DOI: 10.1145/285243.285291. 7, 67, 69, 91, 104, 109, 153

[44] D. Qiu and R. Srikant. Modeling and performance analysis of BitTorrent-like peer-to-peer networks. In *Proc. SIGCOMM*, pages 367–378, 2004. DOI: 10.1145/1015467.1015508. 8, 67, 69, 87, 91, 149

[45] N. Roy, A. Dubey, A. Gokhale, and L. Dowdy. A capacity planning process for performance assurance of component-based distributed systems. In *Proc. ACM/SPEC Int. Conf. Performance Engineering (ICPE)*, pages 259–270, September 2011. DOI: 10.1145/1958746.1958784. 56, 61, 97, 105, 115, 152

[46] B. Schroeder, A. Wierman, and M. Harchol-Balter. Open versus closed: a cautionary tale. In *Proc. Symp. Networked Systems Design and Implementation (NSDI)*, 2006. 8, 56, 58, 86, 152

[47] P. J. Schweitzer. Approximate analysis of multi-class closed networks of queues. In *Proc. Int. Conf. Stochastic Control and Optimization*, pages 25–29, 1979. 46, 153

[48] K. C. Sevcik and I. Mitrani. The distribution of queuing network states at input and output instants. *J. ACM*, 28(2):358–371, 1981. DOI: 10.1145/322248.322257. 44, 149

[49] R. C. Shah, S. Roy, S. Jain, and W. Brunette. Data mules: Modeling a three-tier architecture for sparse sensor networks. In *Proc. IEEE Workshop on Sensor Network Protocols and Applications*, pages 30–41, May 2003. DOI: 10.1109/SNPA.2003.1203354. 40, 68, 93, 110, 153

[50] C. Stewart, T. Kelly, and A. Zhang. Exploiting nonstationarity for performance prediction. In *Proc. Eurosys Conf.*, pages 31–44, March 2007. DOI: 10.1145/1272996.1273002. 90, 98, 151

[51] Y. C. Tay. An approach to analyzing the behavior of some queueing networks. *Oper. Res.*, 40(S2):300–311, May 1992. DOI: /10.1287/opre.40.3.S300. 104

[52] Y. C. Tay. Some performance issues for transactions with firm deadlines. In *Proc. IEEE Real-Time Systems Symposium (RTSS)*, pages 322–331, 1995. DOI: 10.1109/REAL.1995.495221. 111

[53] Y. C. Tay. Data generation for application-specific benchmarking. *PVLDB*, 4(12):1470–1473, 2011. 87

[54] Y. C. Tay. A technique to estimate a system's asymptotic delay and throughput. *Proc. IFIP Performance Conf.*, pages 156–159, 2017. DOI: 10.1145/3199524.3199549. 53

[55] Y. C. Tay and K. C. Chua. A capacity analysis for the IEEE 802.11 MAC protocol. *Wireless Netw.*, 7(2):159–171, 2001. DOI: 10.1023/A:1016637622896. 8, 108, 116, 149

[56] Y. C. Tay, N. Goodman, and R. Suri. Locking performance in centralized databases. *ACM Trans. Database Syst.*, 10(4):415–462, 1985. DOI: 10.1145/4879.4880. 106, 121, 150

[57] Y. C. Tay and H. Pang. Load sharing in distributed multimedia-on-demand systems. *IEEE Trans. Knowl. Data Eng.*, 12(3):410–428, May 2000. DOI: 10.1109/69.846293. 79, 112

[58] Y. C. Tay, R. Suri, and N. Goodman. A mean value performance model for locking in databases: The no-waiting case. *J. ACM*, 32(3):618–651, 1985. DOI: 10.1145/3828.3831. 33, 92, 110, 150

[59] Y. C. Tay, D. N. Tran, E. Y. Liu, W. T. Ooi, and R. Morris. Equilibrium analysis through separation of user and network behavior. *Comput. Netw.*, 52(18):3405–3420, 2008. DOI: 10.1016/j.comnet.2008.09.008. 9, 58, 88, 150

[60] Y. C. Tay, X. Zong, and X. He. An equation-based heap sizing rule. *Perform. Eval.*, 70(11):948–964, Nov. 2013. DOI: 10.1016/j.peva.2013.05.009. 122

[61] K. S. Trivedi, D. S. Kim, A. Roy, and D. Medhi. Dependability and security models. In *Proc. Int. Workshop on Design of Reliable Communication Networks*, pages 11–20, 2009. DOI: 10.1109/DRCN.2009.5340029. 40, 41, 57, 78, 79, 112, 150

[62] Y.-C. Tu, J. Sun, and S. Prabhakar. Performance analysis of a hybrid media streaming system. In *Proc. ACM/SPIE Conf. on Multimedia Computing and Networking (MMCN)*, pages 69–82, January 2004. DOI: 10.1117/12.538806. 11, 31, 70, 77, 92, 110, 151

[63] B. Urgaonkar, G. Pacifici, P. Shenoy, M. Spreitzer, and A. Tantawi. Analytic modeling of multitier Internet applications. *ACM Trans. Web*, 1(1):2, 2007. DOI: 10.1145/1232722.1232724. 2, 6, 46, 56, 61, 77, 86, 104, 109, 151

[64] V. Venkatesan, Y. C. Tay, and Q. Wei. Sizing cleancache allocation for virtual machines' transcendent memory. *IEEE Trans. Comput.*, 65(6):1949–1963, 2016. DOI: 10.1109/TC.2015.2456025. 122

[65] E. Vianna, G. Comarela, T. Pontes, J. M. Almeida, V. A. F. Almeida, K. Wilkinson, H. A. Kuno, and U. Dayal. Analytical performance models for MapReduce workloads. *Int. J. Parallel Program.*, 41(4):495–525, August 2013. DOI: 10.1007/s10766-012-0227-4. 90, 101, 116, 120, 151

[66] S. Williams, A. Waterman, and D. Patterson. Roofline: an insightful visual performance model for multicore architectures. *Commun. ACM*, 52(4):65–76, Apr. 2009. DOI: 10.1145/1498765.1498785. 56, 60, 95, 112, 153

[67] X. Zhang, G. Neglia, J. Kurose, and D. Towsley. Performance modeling of epidemic rout-ing. *Comput. Netw.*, 51(10):2867–2891, July 2007. DOI: 10.1016/j.comnet.2006.11.028. 77, 81, 96, 113, 150

[68] Q. Zhu, Z. Chen, L. Tan, Y. Zhou, K. Keeton, and J. Wilkes. Hibernator: helping disk arrays sleep through the winter. *Proc. ACM Symp. Operating Systems Principles (SOSP)*, 39(5):177–190, 2005. DOI: 10.1145/1095809.1095828. 7, 20, 21, 30, 86, 88, 90, 104, 109, 153

[69] C. C. Zou, W. Gong, and D. Towsley. Code red worm propagation modeling and analysis. In *Proc. ACM Conf. Computer and Communications Security (CCS)*, pages 138–147, 2002. DOI: 10.1145/586110.586130. 7, 77, 79, 91, 104, 109, 149

# Author's Biography

## Y.C. TAY

**Y.C. Tay** received his B.Sc. from the University of Singapore and his Ph.D. from Harvard University. He is a professor in the Department of Mathematics and Department of Computer Science at the National University of Singapore, and a Resident Fellow in Tembusu College. He has spent sabbaticals at Princeton, MIT, Cambridge, UCLA, National Taiwan University, Microsoft, Intel, and VMware. He has served on program committees for ACM SIGMETRICS, ACM SIGMOD, IFIP PERFORMANCE, VLDB, IEEE ICDE, IEEE MASCOTS, IFIP NETWORKING, and ICS. He is also a Senior Associate Editor for *ACM Transactions Modeling and Performance Evaluation of Computing Systems*. His main research interest is performance modeling (database transactions, wireless protocols, Internet traffic, cache misses). Other interests include database systems (synthetic generation of data and social networks) and the use of local time in distributed computing. He has won several teaching awards.

# Index

Printed in the United States
by Baker & Taylor Publisher Services